ALSO BY BYRNE N. SHERWOOD

Twilight in the 'Nam

Reporting For Duty
I loved Every Minute

Reporting For Duty
I loved Every Minute

BYRNE N. "BUZZ" SHERWOOD

Deeds Publishing | Athens

Published by Deeds Publishing in Athens, GA
www.deedspublishing.com

Printed in The United States of America

Cover and interior design by Deeds Publishing

ISBN 978-1-961505-31-5

Books are available in quantity for promotional or premium use. For information, email info@deedspublishing.com.

First Edition, 2024

10 9 8 7 6 5 4 3 2 1

Dedicated to:
Sergeant First Class John Jarrette, Jr.
Platoon Sergeant, 3d Platoon, B Company,
2nd Battalion, 508th Infantry (Airborne),
82d Airborne Division

Lieutenant Colonel (Later Lieutenant General) Neal T. "Tom" Jaco
Commander, 1st/3rd Battalion, 28th Infantry,
1st Infantry Division

Mentors, Comrades, Friends

Sergeant First Class John Jarrette Jr., Platoon Sergeant, 3d Platoon,
B Company, 2d Battalion, 508th Infantry (Airborne)

Lieutenant Colonel (later Lieutenant General) Neal T. "Tom" Jaco,
Commanding Officer, 1st/3d Battalion, 28th Infantry, 1st Infantry Division

Contents

Foreword/Acknowledgements

I was stationed in Germany from 1976 to 1979. This was long before Amazon, Google, Kindle, or any of the other means of procuring information or goods that exist today. In those days, when stationed in a foreign country, the only practical access to English language books was the Stars and Stripes bookstores that were located on most military bases. These stores were quite small and had to cater to the reading tastes of the entire military family. Potboilers, westerns, science fiction, action war novels, and romance novels abounded. Not so much so in the history department. If one was a history buff, as I am, the pickings were scarce.

The upside of this situation was that one was forced out of one's literary comfort zone. In my search for something to read, I stumbled onto gems like *"Eaters of the Dead"*, a marvelous retelling of the Beowulf saga by Michael Crichton.

My biggest find was the Flashman books by George MacDonald Fraser. These were billed as comic novels but were, in fact, well researched historical novels. The only thing comic about them was the central character, Flashman, who, through his roguish adventures carried the reader into various historical events—from the Sepoy Mutiny in India to John Brown's raid at Harper's Ferry and Custer's Last Stand.

In reading the Flashman books, I noted that Fraser had also

written two small volumes titled "*The General Danced at Dawn*" and "*MacAuslan in the Rough*," which were semi-autobiographical stories from Fraser's time as a lieutenant in the Gordon Highlanders in the immediate aftermath of WWII.

Toward the end of my tour of duty in Germany, they appeared in the Stars and Stripes bookstore, and I snatched them up. I loved them, keeping them by my bedside to read again and again whenever I needed a boost. In addition to being entertained, I realized that I, too, had stories that were funny, poignant, sad, and heart-warming. It was here that the seed was planted to someday write a book of my own.

It would take almost fifty years for that seed to come to fruition. Several comrades from various phases of my service have provided commentary and suggestions, but it is to George MacDonald Fraser that I owe the inspiration for this book.

1. Row, Row, Row Your Boat

Row, row, row your boat
Gently down the stream
Merrily, merrily, merrily, merrily
Life is but a dream

Whoever wrote that little ditty wasn't far off, but I don't know about the 'merrily' part.

In 1966, at age seventeen, I became an officer cadet in the ROTC program at UC Berkeley. Early on, I decided that I wanted to be a Ranger. This was the toughest training in the Army, and those who successfully completed the school were held in the highest esteem. Also, since I expected to be sent to Vietnam, this would be the best possible preparation for combat.

Sergeant First Class Silvernail, our instructor for small unit tactics and map reading, was a Ranger, and to me the epitome of the professional soldier. A veteran of two wars with five Purple Hearts to show for it, he was a hard man, while at the same time down to earth and personable. In spite of our greenness, he treated us cadets like the potential leaders he was helping to mold. I wanted to be just like him.

My thirst to become a Ranger was further whetted by the training film about Ranger School. Not only did the training look tough and demanding, it also looked exotic. Rangers did things heretofore unimaginable—fighting each other in the hand to hand combat pit, balance walking on a log forty feet over a lake, rappelling, scaling a sheer cliff face…and the boats. Paddling around in inflatable rubber boats looked really cool and fun.

Four years later, I found myself in Ranger Class 7-71, a winter Ranger. The weather ranged from cold to really cold to extremely cold. By the third and final phase of Ranger School, the jungle phase in Florida, I had done all the activities seen in the film. All, that is, but the boats. That was a feature of the Florida training.

What had seemed exotic when I was seventeen was, in the event, grueling and terrifying—and yet exhilarating. But surely the boats would be fun. How could they not be?

Finally, about halfway through the Florida phase, the opportunity came. We were to paddle down the Yellow River to a specified point and conduct a mission. By this time, I had lost forty pounds and was exhausted from lack of sleep and malnutrition. We averaged 2-3 hours sleep a night and were only being fed one meal a day. I was ready for some fun.

The first clue that we might not be rowing merrily along came when I found out we wouldn't begin until after dark and we would be moving all night. And then there was Sergeant First Class Taylor. I was not airborne qualified at the time and was, along with the other non-airborne students, in the "leg" platoon. SFC Taylor was likewise not airborne qualified and was determined that his "legs" would outshine the airborne guys.

"Alright, you legs, we're going to show these airborne rag bags how it's done. We're going to take the lead and keep it. Is that clear?"

"Yes, sergeant! Rangers lead the way!"

Each boat accommodated nine Rangers, three on each side to paddle, two to navigate and a coxswain who was overall in charge.

Weapons were slung on our backs and rucksacks were lashed down in the middle of the boat.

Navigating a boat at night was an interesting process. One member of the navigation team, the observer, was positioned in the prow of the boat. The other, the navigator, was in the center, crouched under a poncho with a map and flashlight. The observer would whisper each change in the river's course back to the navigator, who would whisper back acknowledgement and plot the progress on the map. This was a tedious, exacting process because one missed turn could cause the group to miss its landing site. Paddling the boat was no picnic either. Paddlers sat on the gunwale with their outboard leg folded under their buttocks. The inboard leg was planted firmly inside for stability. Paddling for several hours in this position became extremely uncomfortable. The coxswain would prescribe the tempo for paddling, announce direction changes and rest stops. Since this was a clandestine patrol, noise was kept to the absolute minimum and all communication was done in whispers.

Sergeant Taylor gave us our final admonition. "OK Rangers, we take the lead right out of the chute and keep it all the way."

"Roger that, sergeant."

With that, the four boats of the "leg" platoon pulled hard and took an early lead.

The rest of the trip was a blur of constant paddling and listening to the whispered exchanges between the observer and the navigator.

"Curving right."

"Roger, curving right."

"Straight on."

"Roger, straight on."

Occasionally, the coxswain would direct a rest stop.

"Pull right."

Upon reaching the riverbank, we were told to back paddle until someone on the starboard side could grab a branch and bring us to a halt. As welcome as the rests were, they never lasted long. As soon as one of the "airborne" boats appeared out of the darkness, we were ordered to press on.

On and on, hour after hour we paddled, the exoticness of it long left behind. Finally, about four a.m., the navigator announced that we were nearing our destination.

"Objective coming up on the left, one hundred meters."

"Roger, one hundred meters on the left."

The coxswain now took over.

"Pull left."

The paddlers on the starboard side continued to paddle while those on the port side dragged their paddles, causing the boat to steer left. Having changed direction, the command came, "All ahead," at which time we all pulled hard for a small beach on the left bank.

After beaching our boat and wading ashore, we discovered that our travails were nowhere near at an end. With our heavy rucksacks still on board, we had to hoist the boat onto our shoulders and carry it quite a distance to a sandy road which ran through the woods. Depositing the boats beside the road, we continued our patrol, moving off in single file.

My ranger buddy, Brian "BC" Campbell, had become delirious from physical and mental fatigue. I grabbed hold of his web gear and guided him along, lest he wander away and get lost. I myself was hallucinating. As the first grayness of dawn began to show itself, I saw a cluster of people, men, women and children, sitting beside the trail brewing coffee over small cook fires. They were dressed in dark clothing, the men unshaven and wearing slouch hats. In general appearance, they reminded me of European gypsies. I was dragging BC along like a puppy not yet trained to the leash.

"Don't talk to them, BC, they're not part of our patrol."

As the new day dawned, we began to regain our senses but, for the life of me, I can't remember what mission we went on to perform. All I know is that we rowed our boats down the stream and our lives, most certainly, became a dream.

"Leg" Boat crew. Front row, kneeling L-R: Brian "BC" Campbell, author. L-R rear: Dave Kuhl, Larry Bliss, Brian Black, Tom Quinn, Bill Namick, Bill Durham, Rich Roecker, Alex Schmidt. Graduation day, Ranger Class 7-71.

2. Reporting For Duty

I was standing at parade rest in front of the desk of Colonel Knulsen, the commander of the First Brigade, 82d Airborne Division.

"OK, Lieutenant Sherwood, I'm assigning you to the Second of the 508. They are located right across the street on your left. Report to Colonel Simmons."

"Yes, sir," I said, coming to attention and snapping him a salute. "All the way, sir!"

Returning my salute, the colonel responded, "Airborne!"

Executing an About Face, I marched out of his office and proceeded to the main entrance, where I paused to get my bearings. The entire 82d Airborne Division was arrayed along Ardennes Street. The two-story brigade headquarters buildings were located on one side of the street and on the other were their three subordinate battalion headquarters in smaller, single story buildings. Behind each battalion headquarters, four large, three-story barracks buildings were lined up, each housing one of the battalion's companies — HQ, A, B and C.

On the left, just as Colonel Knulsen had said, was the 2d Battalion, 508th Airborne Infantry. This was it. My dream was finally coming true. I had been in the Army for eight months, but so far it had all been training: the Infantry Officers Basic Course, Ranger

School, and Airborne School. But now was the pay-off…now I was going to join an outfit and take over the leadership of a platoon of paratroopers.

Stepping off in the direction of the 2-508, I noticed a short, barrel shaped man with gray hair standing just outside the door. He was too far away to see his rank, but this had to be Lieutenant Colonel Simmons. Colonels were old men, and this person had gray hair. Before I even got to the sidewalk, the man disappeared back inside the headquarters, but no problem, I knew who I was looking for.

Intent on my mission, I entered the headquarters, failing to notice the sign by the door, which said, "Commander's Entrance." In front of me was a short hallway, created by partitions on the left, which masked a work area, and on the right were three office entrances. No one was in sight, so I walked along tentatively, looking for some sign of the gray haired, barrel shaped Colonel Simmons.

"WHAT DO YOU WANT?" Someone growled at me from the right. Completely startled, I whirled to the right to find the source of this unseen voice. There, in the darkened doorway, was a small, wiry black man glowering at me.

"I'm here to see Colonel Simmons," I replied.

"I'M COLONEL SIMMONS, SIT DOWN!"

As he spoke, he gestured to the office one door down. Then he disappeared. I went to the indicated office, which turned out to be the office of Captain Leslie Hardy, the battalion Adjutant. I knocked and was greeted by Captain Hardy, who happened also to be black, but there the similarity to Colonel Simmons ended. He was quite friendly and nonchalant. "Take a seat."

"Thank you, sir."

After a little small talk, he asked me some basic questions regarding my marital status, if I had quarters yet, etc. Then he went about his work while I sat waiting and dreading my next encounter with Colonel Simmons. I sat, and I sat. After about two hours, Captain Hardy said, "I don't think the colonel is going to see you today. Go on home and come in tomorrow morning."

"Yes, sir."

"Oh, and hey, that door is the commander's door. Come in through the door on the other side of the building."

I felt a flush rising up the back of my neck as I realized that I had screwed up before even reporting to my unit.

"Yes, sir."

I left feeling a little disconcerted. This wasn't exactly the way I had envisioned it. But here I was, and tomorrow was a new day. Everything would fall into place tomorrow.

I reported to Captain Hardy in the morning and took the same seat as the day before.

"It's going to be a little while," Captain Hardy informed me, "the colonel's in a meeting."

I had been sitting for about an hour when Colonel Simmons suddenly appeared in the doorway bellowing,

"ON YOUR FEEET!"

I popped to attention and stood at a brace, waiting for whatever was going to happen next. Nothing happened. The colonel had left. Captain Hardy had also initially come to attention, but he now went about his business. I, however, remained standing at attention for a good ten to fifteen minutes.

When it became apparent that nothing was going to happen imminently, I cautiously assumed the position of Parade Rest and there I remained for what seemed like a long time.

Without warning, Colonel Simmons was again in our midst, and I sprang back to the position of attention. For some reason, the colonel declined to address me directly even though he, Captain Hardy, and I were no more than an arm's distance apart. Addressing Captain Hardy, he said, "Does he have a wife?"

"Yes, sir, he has a wife."

"Does he have children?"

"I don't know, sir." Then, turning to me, "Do you have any kids?"

"No, sir."

"No, sir, he doesn't have any children."

"Well then, tell him to go home."

"Yes, sir."

And with that, the colonel again vanished.

"OK, Lieutenant Sherwood, take the rest of the day off and come on back in the morning."

Now I was feeling a lot disconcerted, but was at a loss what to do but keep driving on and hoping for the best.

I returned the next morning and was again waiting in the Adjutant's office when Colonel Simmons appeared with somewhat less drama than the preceding two visits. On the wall was a wire diagram of the battalion, showing all the different companies and staff positions. In his hand was a telescoping pointer, which he now extended. "I'm going to make you ..."

As he spoke, he was moving the pointer around on the chart as if it were a Ouija board and he was waiting for the spirits to signal him where to land.

'Yes', I was thinking, 'this is it. A, B, or C company, I don't care. Just get me to a company so I can take over a platoon.'

"I'm, going to make you ..."

The pointer was still traveling the length and breadth of the chart when it abruptly came to a halt.

"The Property Book Officer."

THE PROPERTY BOOK OFFICER!? What the hell was that? I felt the blood drain from my face, and I felt faint. I had no idea what the Property Book Officer was, but I knew for sure that I didn't want to be it.

In a fog, I listened as Captain Hardy gave me instructions to report to Captain Baldwin, the S-4 (Supply Officer), who would be my boss. I was dimly aware that Colonel Simmons had, once again, vanished.

In a state of disbelief that such a calamity had befallen me, I walked down to the S-4's office and reported to Captain Marion Baldwin. Captain Baldwin was a former Special Forces officer with a large moon face which was covered with scars. He spoke with a pronounced stammer. After reporting to him, he said,

"H-h-hey man. T-t-take a few d-d-days off and g-g-get yourself sq-squared away."

Given that I was now extremely disconcerted, this instruction came as a welcome relief. Not only did I need time to digest all this. I also did need the time to find a place to live and get settled in as best as I could in a few days.

When I returned to duty on Monday, I was informed that my assignment had been changed and that I was now going to be assigned to B Company. Hallelujah! I never found out the particulars of the decision to change my assignment. All that mattered to me was that it had been changed.

So, at about 0900, I marched down to B Company. Yes, this was it! I was going to be reporting for duty in my first real assignment. It would be just like in the movies. I would march into the Old Man's office, come crashing to a halt in front of his desk, whip out my best salute, and in a clear, confident voice say, "Lieutenant Sherwood reporting for duty, sir!"

Then this seasoned veteran would give me some sage advice and assign me to a platoon.

Coming into the company orderly room, the first person I met was First Sergeant Rhoden. He was a solid, grizzled looking man who, I could see, had fought in Korea and Vietnam—just the kind of top sergeant you would hope to have in the Army's most elite unit. He indicated that I should proceed to report to Captain Ferguson, the Company Commander.

I stepped up to his office and knocked firmly on the door. "Yeah?"

I opened the door and marched in to report to—a guy sitting behind the desk wearing an electric blue golf shirt and checkerboard slacks—eating potato chips. Standing around the perimeter of his office were the other lieutenants of the company, all wearing the normal fatigue uniform. Undaunted, I made my advance to that invisible spot in front of his desk, halted, saluted and said, "Lieutenant Sherwood reporting for duty, sir!"

What I got in return might be construed as a salute, but it was hard to be sure. "Yeah, hey, glad to have you. This is Lieutenant Hendrickson, Lieutenant Kennedy, Lieutenant Kaiser, Lieutenant Beasley…"

"Buzz, how you doing, man. It's good to see you."

Beas, it was Beas, my buddy from Ranger School. Boy was it good to see a familiar face.

"You two know each other?" Captain Ferguson inquired.

"Yes. Sir, from Ranger School."

"OK, good. Beasley will show you what to do."

It may not have been like in the movies but, for better or worse, I'd finally found my home.

3. Duty Officer

In the Army, someone is always in charge. This is particu-larly true in an elite unit like the 82d Airborne Division, which is always on alert and ready to deploy anywhere in the world within eighteen hours.

At the company level during after-duty hours, the one hundred and fifty or so sleeping soldiers are in the care of the Charge of Quarters, or CQ. The young sergeant who performs this duty is responsible for maintaining the order and discipline of the unit, conducting periodic inspections, and being prepared for the phone to ring, activating the alert process which could send the unit off to war.

At the battalion level, this responsibility falls to the Staff Duty Officer, or SDO, usually a lieutenant, and the Staff Duty Non-commissioned Officer, or SDNCO, usually a Staff Sergeant. Also on hand would be a private soldier who would serve as a runner. Selection for this duty is done via the Duty Roster, which ensures that all eligible officers and NCO's are systematically selected in a fair and impartial way.

In the 2d Battalion, 508th Airborne Infantry, there was indeed such a duty roster, but it was used as the exception to the rule rather than the rule. The Battalion Commander, Lieutenant Colonel

Simmons, more commonly referred to as "Colonel Smoke"[1], used the role of SDO as a punishment for officers who, through some infraction, incurred his wrath. Some would have described Smoke as wrath incarnate and made every effort to steer clear and stay off his radar. Not so Lieutenant Collins.

Lieutenant "Fast Eddie" Collins seemed to live in Smoke's radar and, as a consequence, was like the permanent duty officer. Fast Eddie was, according to one of my sergeants who had known him then, a small time hoodlum from Cleveland. Somehow he ended up being commissioned as an officer, but came up short at being the gentleman that officers were supposed to be. Amongst other infractions, Fast Eddie didn't subscribe to the physical image of the U.S. paratrooper, striving rather for his own version of olive green sartorial splendor.

He wore his hair extremely long by Airborne standards, effecting as closely as possible the swept over the ears look favored by tel-evangelists. He had developed his own style of wearing his fatigue uniform, tacking the points of his collars down so that it resembled a button-down collar, and turning the cuffs of his shirt under one time and ironing them down. In cooler weather, he topped the ensemble with an aviator's flight jacket. All of this was highly unorthodox and highly unacceptable to Colonel Smoke.

The impact of Fast Eddie on my life was that it was quite a while before I pulled SDO for the first time, but the day inevitably came. I reported to the Adjutant at 1600 hours for my briefing

---

1. In the common Army slang of the day, to "bring smoke" on your enemy in combat was to bring overwhelming death and destruction on them. In inter-personal relations, to "bring smoke" on another person was to devastate the other by non-lethal means such as verbal abuse, punishment, reprimand, etc. Colonel Simmons was a master at "bringing smoke" down on his subordinates, hence his nickname.

and special instructions and, having received them, was sitting in Captain Hardy's office waiting for the end of the duty day and the beginning of my stint as SDO. Captain Hardy left the office for a few minutes and then re-appeared, announcing to me, "The colonel wants to see you."

'Oh shit!' I thought, 'Have I done something wrong? What does he want to see me for?'

Right or wrong, the thought of being in a one on one setting with Colonel Smoke was nothing short of terrifying. I stood up, stepped to the doorway to his office, knocked on the door-frame, and said, "Sir, you wanted to see me?"

"Come in here."

Stepping up next to his desk, he said to me, "Can you draw?"

A voice inside of me was screaming, 'Say no! Don't put yourself in the cross hairs. Don't do it!' Then I heard myself saying, "Yes, sir, I can draw."

The next thing I knew, he was pushing a note-pad in front of me and handing me a felt-tip pen, saying, "Draw that!"

Looking up, I saw that he was pointing to a photograph of President John F. Kennedy hanging on the wall in front of his desk. Full of alarm at being put on the spot like this and fearing what would happen if I failed to satisfy, I feverishly began drawing.

After a couple of minutes the colonel said, "OK. That's good. Here's what I want you to do. I'm briefing the brigade commander tomorrow on our training program. The briefing charts are all completed, but I want you to draw a red devil on the top of each chart and then print 'RED DEVILS.' Can you do that?"

"Yes, sir."

"Captain Hardy will get you the charts and the markers. I want it to be good."

"Yes, sir."

Given this mission that would probably take me all night, I was barely aware of Staff Sergeant Sneed and a brand new private showing up to assume the duties of SDNCO and runner.

Except for me working frantically through the night to finish my special task before morning, the night was seemingly uneventful. Pre-occupied as I was with my work, I pretty much left Staff Sergeant Sneed alone to supervise the cleaning of the battalion headquarters. The runner, when not otherwise engaged, sat at a desk spit-shining his paratrooper boots, which he had gleaming like glass.

At 0430 hours, in came Command Sergeant Major Walter Sabalauski, the white haired, barrel-shaped man whom I had mistaken as the battalion commander on the day I reported for duty in the battalion.

CSM Sabalauski, known as "Sab" or "Sabo" was a legend in the Army. A professional boxer before entering the Army prior to WWII, he was on Pearl Harbor when it was bombed, went through WWII in the Pacific, the Korean War, and three years in Vietnam. He was the recipient of the Distinguished Service Cross, the second highest medal for heroism that the Army bestows.

His solution to any problem was to fight. At sixty-one years of age, he struck terror into the hearts of all the NCO's. In fact, as I was about to find out, he could strike terror into the hearts of anyone.

"God damn it. We're going to have fucking majors start pulling duty officer." He screamed, throwing a push-broom at the startled Staff Sergeant Sneed. The runner, caught mid-stroke in his spit shining, sat with his cotton ball hovering above the brilliantly shining toe-cap of his boot.

"Put those in the garbage!" He screamed at the runner, who stared, immobile, saucer-eyed and uncomprehending.

"I said put the fucking boots in the garbage can!"

Yanked from his stupor, he dropped the boots into the trash-can as if they were red-hot.

The rest of the morning went by in a blur of being busy, trying to look busy, and trying not to attract the attention of the CSM. At 0600 hours as the normal duty day started, I dutifully handed over the briefing charts to Captain Hardy and was excused. I never heard anything more about the night, so I assumed that the charts were acceptable and that the fury of the CSM had no lingering ef-fects. In time, I got to see other sides of the CSM and, when I was promoted to First Lieutenant, he hugged me as if I was his son.

All in all, it was a night not soon to be forgotten. Forty-five years later, it still makes me shudder.

Command Sergeant Major Walter J. Sabalauski

4. May Day 1971

CLOSE THE PENTAGON

"Sergeant Washington, can I get you another beer?"

"Yes, sir, that'd be great."

The can made a '*pfffhhht*' sound as I popped the top and handed it to him.

"This chicken is good. Where did you learn how to cook, sir?"

"Hey, after a few beers, everything tastes good."

It was a warm North Carolina afternoon on Sunday, May 3, 1971. My friend Lieutenant John Beasley and I were hosting a barbecue at my house for the sergeants of Weapons Platoon, B Company, 2d Battalion, 508th Airborne Infantry, 82d Airborne Division.

John, two months senior to me, was the overall leader of the platoon, with specific responsibility for the mortars, while I had responsibility for the anti-tank section. Sergeant First Class Denny, an African American from West Virginia, was the platoon sergeant. Staff Sergeant Reynolds, African American and recently returned from his second tour in Vietnam, was the mortar section sergeant. Sergeants Walker, Washington and Jenkins were squad leaders. Staff Sergeant David Beran was the section sergeant of the anti-tank section, with Sergeants Hosey and Eveland as squad leaders.

As we sat there in my backyard under the ubiquitous pine trees, we chatted about what we had been up to and what we expected to be doing.

"I'll be glad when we get off post-support."

"You know that's right! Especially burial detail. Remember when we had to bury that guy way out in the boonies? It was raining and they buried him up on the side of a hill. We had to carry the casket up the hill, and we were slipping all over the place in the mud."

John piped up, "How about the other day when we were all sitting in the CP (Command Post)[2] and the burial detail was practicing just outside the window. When the firing squad fired a volley, everyone except Lieutenant Sherwood and me was down on the floor. It was pretty funny."

"Hey, sir," said Sergeant Reynolds, "I just got back from Vietnam a month ago. You hear a shot, you go down."

All the other NCO's nodded in agreement.

"I know, I know," said John, "it just seemed funny that we were all sitting there talking and then, all of a sudden, it was just the two of us."

"How was it over there?," I asked Sergeant Reynolds. Sergeant Reynolds was unusual for an NCO, a bespectacled, professorial type who spoke very deliberately and correctly, unlike most people in the Army, who spoke in regional dialects, used a lot of slang and cursed profusely.

"It was very disappointing, sir. Not at all like the first trip. I don't think I care to go back."

"How do you mean?"

"Oh, the morale wasn't as good, even in The Herd (the 173d

---

2. Command Post — the euphemistic term for the platoon office

Airborne Brigade), there were too many restrictions about what you could and couldn't do. It just wasn't the same."

"Well, you know right after we get off Post-support we'll be going to Indiantown Gap for ROTC Summer Camp."

At the mention of Indiantown Gap, Sergeant Hosey brightened noticeably. "Hey, that's what I'm talkin' about. Hershey P-A here I come. I'm lookin' forward to seein' them girls from the Job Corps. Last time we were there, they had dances for us at the Job Corps center."

This produced chuckles from the other NCO's. I was smiling inwardly as I tried to imagine what girls at a Job Corps Center looked like.

Rrrring, rrring.

"Hang on, let me catch the phone."

This was before the days of answering machines, pagers, or cell phones. The telephone was the only link with the rest of the world, and one always answered the phone. I picked it up.

"Lieutenant Sherwood's quarters, Lieutenant Sherwood speaking."

From the other end came the most unexpected news. "Sir, Sergeant Carter from the company. We've been alerted. You need to activate the telephonic alert and report back ASAP."

Stunned, I responded, "Roger that. I've got all the NCO's here. I'll be right in." I hung up.

Normally, my next action would be to telephone all the NCO's under my supervision and they, in turn, would notify any of their soldiers who did not live in the barracks. But, since they were all at my house, that was unnecessary. I hurried out to the assembled guests.

"We've been alerted. Get back to the company ASAP. I'll see you there."

They looked as surprised as I was, but we were all used to responding quickly to the unexpected. That we were surprised was an indication of how little attention we paid to what was going on outside our military world.

After everyone had hurriedly left, I changed into my fatigues and packed my field gear and extra fatigues into my duffle bag. I was excited. In the blink of an eye, we had gone from the boring predictability of Post-support to being alerted to go to who knew where. My wife, on the other hand, was terrified. "What's happened? Where are you going? How long will you be gone?"

"I have no idea. I'll call you if I can, to let you know what's going on."

John and I arrived at the company at about the same time and were the first officers to get there. The NCO's who lived in the barracks were already there and those who lived off-post soon arrived. Being on Post-support, our battalion was on the lowest level of alert status, so things were rather chaotic in the company. The soldiers in the barracks were busy packing their gear and, as others trickled in, they were galvanized into action. John and I reported our presence to battalion as well as the number of soldiers we had on hand.

It should be mentioned at this point that, by 1971, the Army had been fighting an unpopular war in Vietnam for six years and was in bad shape. The 82nd Airborne Division was the elite formation in the Army and was in much better shape than the rest of the Army, but it, too, had suffered. Hence, the relatively disorganized response to the alert. And, we still didn't know why we had been alerted.

"Where are we going?"

"Hell if I know."

"I heard we're going to Puerto Rico."

"Puerto Rico! Why would we go to Puerto Rico?"

"I dunno, don't they want to be their own country? Or become a state? Something like that."

The idea of going to Puerto Rico took hold and there was much talk about going there and what our mission might be.

We were confined to barracks. Once arrived, no one was allowed to leave.

John and I, who seemed to be joined at the hip as co-leaders of whatever was going to happen, were summoned to battalion headquarters, along with representatives of the other companies. We now learned what was going on.

"Gentlemen, there is a massive demonstration in Washington D.C. which seems to have the aim of closing the Pentagon. The federal authorities have activated Garden Plot (the code word for the activation of the U.S. military to quell civil disturbance). Everyone is confined to barracks. Be prepared to deploy on a moment's notice. The uniform will be fatigues with field jacket, LBE, protective mask, steel pot and M-16 with bayonet. No machine guns or heavy weapons are to be deployed. We will keep you informed as the situation develops. Are there any questions?"

There being none, everyone returned to their companies to prepare and await developments. Meanwhile, Captain Ferguson and First Sergeant Rhoden had arrived to take charge of the whole company, so John and I withdrew to the lesser realm of Weapons Platoon, inspecting equipment and awaiting developments.

Keeping many thousands of soldiers confined to barracks amidst the electricity of an alert is no small thing, and I'm sure the senior commanders at division were pulling their hair out. Some relief came from an unexpected source. How or why they happened to be available, I have no idea, but what I do know is the "Dixie Cups", a female soul group from New Orleans who were very popular in the early 60's, suddenly appeared on the scene and, after dinner in the mess hall, we were ordered to march our soldiers down to York Theater for a concert.

Company B had assembled an ad hoc platoon made up of members of all platoons in the company with John and I in charge, and we dutifully joined the columns marching to the theater. As hundreds of soldiers converged on the theater, military organization prevented chaos. As we arrived, we remained in formation and waited for our turn to file in and be seated together as a unit.

Once the theater was packed to capacity, the lights were dimmed, and we waited for the show. Then, out they came, The Dixie Cups. Many or even most of the younger soldiers would have never heard of them. Music tastes had moved from Rhythm and Blues and Oldie Rock and Roll to psychedelic music and harder, more strident themes of protest. But the Dixie Cups were nothing if they weren't entertainers and they soon had the audience in the palms of their hands.

They began with their number one, signature hit, "Chapel of Love."

Gee I really love you
And we're gonna get married
Goin' to the chapel of love

You'd have thought we were back in 1964, listening to the Billboard number one hit. The crowd loved it. As they moved on to some of their livelier numbers, the audience really came alive. Iko Iko really brought them to their feet, literally.

My grandma and your grandma were sittin' by the fire
My grandma told your grandma, "I'm gonna set your flag on fire

Talkin' 'bout hey now, hey now I-ko I-ko un day Jock-o-mo fee-na ai na ne, jock-o-mo fee na ne

First, one black soldier jumped up on the stage and started dancing, followed by several more. Other soldiers leaped to their feet and were dancing by their seat or in the aisle. It was wild and exuberant, and I loved the music, just as I had in its day. But not being one given to spontaneous demonstration, I felt totally out of place and was happy when the concert was over. We exited as we had come in, in an orderly fashion, forming up outside the theater and marching back to barracks.

Once back and the troops settled in, John and I reported to Captain Ferguson for the latest update. In the morning, we were to take our composite platoon to the 2d Battalion, 325th Airborne Infantry, to provide fillers for them as they were short of their full complement of soldiers. After briefing Sergeant First Class Denny and the section sergeants, we turned in to get what sleep we could, sleeping on the floor in the CP.

First thing in the morning, John and I went down to the headquarters of the Second of the Three-Two-Five for instructions, where we received a warm welcome. Unit rivalries became secondary to their need to bring their unit up to strength and our desire to get in on the action as soon as possible, even if it was with another outfit. We were told we would remain an intact platoon and would be assigned to their B Company, which would bring them up to full strength.

When we got back to the company, SFC Denny had the platoon ready to go. We shouldered our duffle bags and rifles and marched up to our new temporary home. Settling the troops in the company day room, John and I, along with the other platoon leaders, reported to the company commander for orders.

"Here's the situation: as you know, there's a major protest going on in Washington D.C. whose aim is to close the Pentagon. Local law enforcement is tapped out and they have activated Garden Plot. Units of the division are already deploying from here to Andrews Air Force Base, where we will assemble and await further orders. We will be moving to Green Ramp[3] later this morning to wait our turn. Be ready to move at any time.

It wasn't long before we got the word. Cattle car trucks were lined up on Ardennes Street to take us down to Green Ramp, which was about a ten-minute drive. Upon arrival at Green Ramp, we were organized into planeloads, or chalks, with the senior officer in each chalk designated as the commander. Green Ramp consisted of long, concrete pads, or ramps, where we would sit while waiting to board planes for a jump or, as in this case, a deployment flight.

We were there for many hours. To the cynic, it seemed like the typical hurry up and wait that one associates with the Army. To me, though, it was obvious there would be waiting involved when moving thousands of soldiers.

Finally, in late afternoon, our turn came to board C -130 cargo planes for the trip to Andrews. The actual flight only took about an hour. Under grey, drizzly skies and unseasonably cool weather, we disembarked the plane and were led to a huge area covered by GP-Large tents full of canvas and wood Army cots. SFC Denny and the other NCO's got the troops situated. With a little bit of quiet time, we started pumping SFC Denny for information.

"Sergeant Denny, didn't I hear you say once that you got deployed to Detroit in '68?"

---

3. Green Ramp was a point where Fort Bragg and Pope Air Force Base adjoined and from which we would get on planes for a parachute jump or a deployment.

"Yes, sir."

"What was it like?"

"Shit, sir, it was like World War Two. I mean, we'd be goin' down a street and snipers would open up on us from the buildings and we'd light 'em up. Those jeep mounted fifty cals would just light 'em up. They would fire up the whole building, and I'm here to tell you a fifty cal brings pee. It would just tear up the whole building. A lot of people got killed."

"What was it like for you?"

"I don't mind telling you, I was scared shitless. It was as bad as Vietnam. I don't want to go through that again. No, sir, buddy."

Well this was sobering, but exciting as well. I had only been in the unit for two months and most of that time had been spent on Post Support. I'd been to the range a couple of times to shoot the anti-tank guns and done a little training, but nothing exciting. Now here we were on a real deployment. There was no philosophical thought about taking to the streets against fellow citizens. We were soldiers and we did as ordered. If the rioters were designated to be our enemies, then that's what they were. We were their enemy as well, I suppose. In any event, I didn't give it a second thought.

Mess tents had already been set up so we were able to get a decent meal for dinner, after which we repaired to our tent to wonder what the morrow would bring.

In the morning, after a hearty Army breakfast, the company commander called us together to brief us on the situation.

"At present, we're being held in reserve. We'll only be deployed to the streets if the police and other local law enforcement agencies can't control the situation. In the event that we are deployed, these are the rules of engagement: 'Ammunition will not be issued unless we are pinned down and taking casualties.'"

I gasped inwardly. *"Pinned down and taking casualties before we will be issued ammunition? That's crazy! Who came up with that?"* I thought.

The company commander continued his briefing.

"We'll begin riot control training at 0900. Formation here in the company street will be at 0845. Uniform will be fatigues, field jacket, steel pot, LBE, protective mask, and weapon with bayonet. Platoon sergeants will report to the First Sergeant for more details. Any questions?"

There being none, we were dismissed to go back and brief our platoons. When we got to the part about rules of engagement, SFC Denny, undoubtedly thinking of his experience in Detroit, burst out,

"What!? Oh fuck that, sir. That's bullshit."

"Hey, Sergeant Denny, that decision comes from way above our pay grade."

Being a professional NCO, he quickly regained control of himself, but he was clearly unhappy about it, as were the rest of us. We were soldiers trained for war, and the idea of going to war unarmed was difficult to reconcile. But, because we were soldiers, we accepted our orders, whether we liked them or not.

At 0845 we fell out to begin our riot control training. Neither John nor I had ever done any riot control training, and we were most eager to learn as much as we could before being deployed to the streets.

First we learned the overall strategy, which was to break up the crowds and move them away from critical areas and disperse them. It was imperative to move crowds along routes that would allow them an avenue of escape. Never were we to box crowds in where they would have no means of escape. The tactic to accomplish this at the unit level was reminiscent of 18th Century tactics rather than the open, dispersed tactics of modern warfare. The basic formation was a line or an inverted V, or wedge, with soldiers marching shoulder to shoulder at the half-step, jabbing the rifle with fixed bayonet every time the left foot hit the ground. If being pummeled by projectiles and spit, this took iron discipline on the part of the soldiers and put tremendous strain on the officers to maintain control.

The more we practiced, the better we got, and the better we got, the more we hoped we would be sent to the streets. In our young minds, there was a game on, and we did not want to sit it out on the sidelines. It was like, "Hey, coach, put me in."

It looked like our wish was going to come true when we were loaded on civilian buses and driven into Washington D.C. and parked adjacent to the Pentagon. From there, we could be rapidly deployed if it became necessary. While waiting there, several of the Pentagon brass came out to thank us for being there. That expression of appreciation really pumped us up.

In the end, we weren't sent in. Owing much to the unseasonably cold weather and drizzle, the crowds were not as immense as had been anticipated and the local law enforcement agencies were able to contain the situation.

In spite of the disappointment, it was still an adventure from which John and I learned a great deal.

Epilogue: In light of recent civil disturbances throughout the country, I hope this story will help the reader to understand the way young soldiers (or police) think in these situations. There is no time for philosophical thinking or introspection. One thinks about what has to be done in the moment, and nothing more.

And, for young, aggressive men, identifying fellow citizens as the enemy is easy if that is whom you are to be pitted against. Which suggests that, when employing young men in that fashion, the thinking of wise elders must prevail. In this case, it prevailed on two counts. First, the decision to hold us back unless it was absolutely necessary and, second, to prescribe the very restrictive rules of engagement under which we were to operate. Whoever made those decisions understood the gravity of deploying soldiers against the citizenry and that, if that grave decision had to be made, every effort must be made to ensure soldiers didn't spill the blood of their fellow citizens.

5. A World Apart

WILKES AND WATAUGA COUNTIES

In mid-September 1971, I took leave to go home and see my parents in California. I hadn't been home since Christmas of 1969 and had subsequently only seen them briefly when they came to New Orleans for my wedding in December of 1970.

While home, I dropped a heavy footlocker on my bare foot, producing a compound fracture of my right big toe. Arriving back at Fort Bragg in a walking cast, I was dismayed to discover that the company had deployed to the field for at least three weeks. Prior to my taking leave, Captain Ferguson had said nothing about the company going anywhere, which was why I decided it was a good time to go on leave.

"What!? They're in the field? Captain Ferguson never said anything about going to the field!!"

"Well, sir, you know how it goes."

Captain Ferguson was grossly incompetent. It was entirely possible that he had known of this deployment all along and never bothered to tell anyone.

"Where'd they go?"

"They're up in the mountains, by Boone, playing aggressor against Special Forces."

"I gotta get up there."

The first thing I did was beg to have the cast removed, which was done in exchange for a set of crutches. After a couple of days hobbling around on crutches, I reported to the battalion surgeon and requested that he clear me for field duty.

"The wound is healing well, but it's going to be tender for a while. Make sure you keep it clean and dry. Your medic can change the dressing for you as needed, and remove the stitches when they're ready to come out."

"Roger that, sir."

Now that I was cleared, I needed to find a way to get up there. Back at the company, I discovered that a sergeant from another platoon was going up there the following day and would give me a ride.

Meeting up with him the following morning, I was surprised to find that we were traveling in his personal car.

'We're going in your POV?"

"Yes, sir. Captain Ferguson let everyone take their POV's."

"You must be shitting me."

"No, sir, I'm not and it's been a mess. There've been all kinds of wrecks and run-ins with the locals. When the company first got up there, they were bivouacked half way up some mountain and the local people started shooting at them."

"No shit?"

This was beginning to sound exciting, something totally different than the formulaic tactical exercises I was used to. I couldn't wait to get up there.

"So where, exactly, are we going?"

"The operation is in Wilkes and Watauga counties, right where North Carolina, Virginia and Tennessee come together. Boone is the biggest town in the area. It's in the Blue Ridge Mountains and is definitely hillbilly country. We're playing the role of the aggressor against the Special Forces. It's completely a free-play exercise."

"What does that mean?"

"There's no boundaries and no restrictions. The local population are even part of it. The SF recruits locals to work for them. We recruit locals to work for us. It's a wild scene."

Upon arrival at the company CP (Command Post), Captain Ferguson was nowhere to be found. I learned that each platoon had its own sector to guard and patrol, so I managed to hitch a ride to the 3d Platoon where I was greeted by my platoon sergeant, SFC John Jarrette.

"Welcome back, sir. What did you do to your foot?"

After explaining about my accident, I asked Sergeant Jarrette to brief me on the situation.

"We're playing the role of the army of an enemy country. The SF are trying to organize an insurgency against the government. Our role is to protect the local government and to conduct counter-insurgency operations. I've placed a squad on the most important bridge in our sector and a fire team on two others. We have a reaction force here in the CP area, as well as a Gama Goat and jeep for transportation. Sir, if you want, I'll show you around the area."

Sergeant Jarrette had set up the bivouac in textbook fashion. He selected a flat area next to a creek, which would provide level ground for sleeping, with water for drinking and washing. The latrine was dug downstream and uphill from the bivouac, so as to prevent contamination of the water supply. Rations and equipment were neatly stored.

"This is outstanding. Great job. So tell me, what's all this about the troops getting shot at?"

"Oh, sir, that was a trip! We got up here on Friday and set up a company bivouac half way up this mountain called Tater Mountain. Since it was Friday, Captain Ferguson released the company for the weekend..."

"He what!? He released the troops out in the middle of nowhere?"

"Yes, sir."

"Unbelievable!"

"So, anyway, all the troops are streaming down the mountain, heading for Boone, some in POV's, others, walking. This pickup truck comes along and picks up a few troops who were walking. The people are named Eisenhower, and they live on top of the mountain. Supposedly they are like outlaws. While they were going down the mountain, the pickup got a flat tire and the Eisenhower's told the soldiers to fix it. Being typical GI's, they said 'fuck you' and kept on walking. The Eisenhower's pulled out shotguns and started shooting at them."

"You gotta be kidding! Was anyone hit?"

"No sir, they ran like hell, though."

"Then what happened?"

"Well, it got reported to the police and they came out. They had sub-machine guns in the trunk of their car and everything, but they were afraid to go up there against these people. In fact, they wanted us to go. Of course, we couldn't go, especially since we only have blank ammo. So, in the end, nothing happened."

"Jesus Christ! Anything else?"

"Let's see. A Deuce-and-a-half full of troops went off a cliff, but the trees stopped it and no one got hurt.[4] And, a combined team of revenuers from North Carolina, South Carolina, and Georgia

---

4. I found out later that it was Sergeant Jarrette who got on the downhill side of the truck, which was only being held up by some small saplings, to extricate the soldiers trapped inside. I put him in for a Soldiers Medal, which he received.

arrived here at the same time as us to go after moonshiners. The locals assumed we were here to assist the revenuers and were very hostile toward us at first. We've been working hard to convince them that we aren't with the revenuers. Now they want to buy our weapons. They are offering a thousand dollars for an M-60."

"What fucking planet are we on? Is there a lot of moonshine?"

"You wouldn't believe it. There's moonshine everywhere. Apparently the stills are huge."

As if to reinforce what Sergeant Jarrette had been telling me, the farmer on whose land we were staying showed up with a jar of moonshine.

"I just thought I'd bring you boys a little sample of some good stuff. They's a lot of bad stuff out there, but I know where this comes from, and I can vouch for it. An' tell yer boys to be careful, the bad stuff can make you go blind or even kill you."

"So how do you tell what's good and what's bad?"

"Fust thing you do is hold it up to the light. If it's crystal clear, it's good. If it's cloudy, it's bad."

"OK, thanks very much."

After the farmer left, I took a small sip of the moonshine he had left. It tasted like all other moonshine I had ever tasted — horrible.

"Oh man, this shit is awful. I don't see how anyone can drink this."

"Well, sir, there's no accounting for taste."

"How about let's take a ride and you can show me around our AO (Area of Operations)?"

As we drove along in the jeep, I asked Sergeant Jarrette to bring me up to date on the operation against the Special Forces.

"So far, sir, we've been kicking their ass. We guessed right on their DZ (Drop Zone) and were there when they jumped in."

"Was it a night jump?"

"Yes, sir. They managed to get away, but we captured all their heavy drop—rations, equipment, etcetera. They're going to be getting hungry so we can be looking for them to be trying to contact the locals for food. We've identified a house that may be a safe house for them. I've posted an OP (Observation Post) there to keep an eye on it."

"What are they trying to accomplish and what is our mission?"

"Well, what they briefed us on is that the SF want to seize control of Boone—you know, city government offices, police station, radio station, post office. Before they do that, they have to establish a support network among the locals. Also, they will probably want to capture and destroy the bridges in order to prevent any government reinforcements from getting here. Our mission is to guard the bridges and to disrupt them from organizing the locals. Our AO is about twenty square miles. The other platoons in the company have similar AO's and missions."

I was having a mixture of reactions to this information. I was extremely impressed with the skill with which Sergeant Jarrette had deployed the platoon and with the results he had achieved—but I was also a little intimidated. I wasn't so sure that I would have been so clever had I been on the scene. On the other hand, Sergeant Jarrette was the consummate professional, making no attempt to steal the spotlight and quite happy to hand the reins over to me. I was also quickly getting caught up in the challenge of defeating the famous Green Berets.

Unlike most military exercises, which were mostly scripted, this was a free play exercise pitting our wits and skill against those of the vaunted Special Forces. And so far, it seemed as if we were doing pretty well. I became more and more drawn into the notion that we could and would decisively defeat them.

I was shaken from my reverie as we arrived at the first bridge that was being guarded by the platoon. While not a huge bridge, it was on the outskirts of the city of Boone and controlled vehicular access to the town. Guarding it was 2nd Squad, under the leadership of Staff Sergeant Nathan Carter. Sergeant Carter was a taciturn individual, not given to easy banter, but highly competent and reliable. He had been the platoon sergeant before Sergeant Jarrette, and I always suspected he was somewhat chagrinned that he had been displaced due to Sergeant Jarrette outranking him. He was thoroughly professional, though, and any resentment he felt was kept in check.

"Hi, Sergeant Carter, how's everything goin'?"

"All The Way, sir.

"Airborne," I replied, to complete the traditional greeting of the 82nd Airborne Division.

"Welcome back. Everything is going good here. We've got a nice, cozy place to bivouac over here in the woods and the locals are always bringing us food. We've hardly had to touch our C-rations. We conduct patrols up and down the stream, looking for any signs of infiltration and I keep a guard on the bridge at all times."

"Outstanding. What can I do for you?"

"Nothing, sir. We're in excellent shape here."

"OK. If anything happens, get us on the horn ASAP and we'll be here with the quick reaction force."

"Roger that, sir."

Moving on, we headed out into the country, on our way to the next bridge, which was being guarded by a four man fire team from 1st Squad, under Sergeant Parker. This was beautiful country, very mountainous and covered in hardwood trees, which were changing color with the onset of Autumn. The weather was warm and sunny. At a bend in the road, we pulled off next to a small bridge over a creek. Specialist Delgado was on guard and, as soon as we pulled up, Sergeant Parker emerged from the woods.

"All the Way, sir!"

"Airborne! Hey, Sergeant Parker. How's everything going here?"

"This place is unbelievable, sir. A little while ago, this guy pulls up and hands me a brown paper bag out of the window of his car. Inside was a jar of moonshine. Also, he offered to pay a thousand dollars for an M-60 machine gun."

"You gotta be joking! Lemme see the moonshine."

Disappearing back into the woods for a moment, Sergeant Parker came back carrying a mason jar. He handed it to me, and I held it up to the light.

"You see how cloudy it is? You can hardly see through it. This is bad shit. The guy who owns the farm where we are bivouacked told me how to tell the good stuff from the bad. This shit can make you go blind."

With that, I poured it out. Now that I was an expert at distinguishing good moonshine from bad, I left Sergeant Parker with some words of caution.

"Be careful with this stuff. Hold it up to the light. If it isn't crystal clear, it's bad. Don't drink it. And listen, if they're willing to pay a thousand bucks for an M-60, they'll be willing to steal one. Take extra precautions with your weapons security—no weapons ever left unattended."

"Airborne, sir."

As we continued on, I remarked to Sergeant Jarrette, "Wow, this is going to be a challenge. It's not even a case of keeping the booze out of the field, it's more a case of trying to keep them away from bad booze."

"I'll put out the word to the NCO's. They'll keep everything under control."

As we continued out into the country, we entered a small valley which was surrounded by wooded hills. On our right was a house, set back, with a large pond between it and the road. Speaking to our driver, who was also my radio operator, Sergeant Jarrette said, "Slow down, Hartsfeld."

As he slowed, Sergeant Jarrette continued, addressing me, "See that house? That's the house I told you about that we suspect is a safe house for the SF. Look about half way up that hill on the left, across the road, and you can see our OP."

Sure enough, as I looked carefully, I could see the two-man OP, looking for all intents and purposes like cattle grazing on the hillside.

"Do they have a radio?"

"Yes, sir. They call in regular SITREPS and, if they see anything, they'll let us know and we can come out with the reaction force."

Continuing on, we came to a bridge over a wide stream. Just before the bridge there was a dirt road leading down to the stream bank below. Before turning off, I could see PFC Lionel Mitchell walking guard at the far end of the bridge. Just past the bridge, the road wound steeply up a heavily wooded mountain. Down below, we were greeted by Sergeant James Hosey, who was guarding this bridge with his five-man fire team. Sergeant Hosey was a blond haired, blue-eyed Georgian, easy going, but not a self-starter. Being a fire team leader under a strong squad leader was the right place for him.

"All the Way, sir! How's it goin'?"

"I broke my toe while I was on leave, but I'm back in the saddle. How is everything here?"

"This place is a trip, sir. Yesterday these two girls came down off the mountain. This one says to the other one, 'Look, them's marryin' stock.'"

"You gotta be kiddin' me."

"I'm not, sir, and I think they were serious."

"So then what happened?"

"Well they just hung around for a while and talked to us. Then they left."

"What'd they look like?"

"One wasn't bad lookin'. The other one looked like a sack of potatoes wearing high top tennis shoes."

We stayed for a while, relaxing in the shadow of the bridge on the broad stream bank where the men were camped and then headed back to the platoon CP, where we kept the Quick Reaction Force (QRF). The QRF consisted of 3rd Squad, led by Sergeant Ronald Carter and the Weapons Squad, led by Staff Sergeant David Beran. Sergeant Ronald Carter shared with Staff Sergeant Nathan Carter the same last name, and both were African American, but aside from that they couldn't have been more different in personality.

Unlike the reserved, business-like Nathan Carter, Sergeant Ronald Carter was outgoing, always smiling, and, in spite of his slender frame, was a bottomless pit when it came to eating. Staff Sergeant Beran was considered the ne'er do well of his family, having enlisted in the Army as a private against the wishes of his father, who was a full colonel in the Air Force. He was intelligent, friendly, and easy going, but not highly motivated.

My original plan had been to rotate the deployed squads back to the CP, to be replaced by one of the QRC squads, but the deployed squad leaders begged to be left in place. Each was having their own unique Blue Ridge Mountain experience and didn't want to give it up for the relative comfort of the CP. Seeing no need to force the issue, and after discussing it with Sergeant Jarrette, I decided to leave things as they were.

YOU GOTTA HELP ME SARGE

After a couple of days of negative SITREPS from our OP, we got the call we had been hoping for.

"Bravo Three-three, this is OP. We've been observing a lot of activity. Several vehicles have been coming and going and there's a couple of males at the house that we haven't observed before."

"Roger. We're on our way. We should be at your location in one zero mikes."

"Sergeant Carter, Sergeant Beran! Saddle up, let's go.

It only took a couple of minutes for the soldiers to put their gear on and load up on the truck. The two squad leaders jumped into the jeep with Sergeant Jarrette and me. As we drove, I gave instructions. "Sergeant Carter, when we get there, take your squad around behind the house and seal it off from anyone leaving or trying to come in. Apprehend any military age males coming or going. Sergeant Beran, you seal off the front of the house and be on the lookout for the two males reported by the OP."

"Yes, sir," they responded.

As we entered the narrow valley, the road bent to the right, giving a clear view of the house. It was set back from the road about twenty-five meters, with a large pond between the house and the road. A semi-circular drive cut between the house and the pond. We pulled into the drive and came to a screeching halt. Sergeant Carter's squad jumped from the truck and followed him toward the back of the house. Sergeant Beran's squad likewise dismounted and started making their way toward the front of the house, Sergeant Jarrette and I following behind.

As we approached, we could see two men sitting at the near edge of the pond, fishing. Sergeant Beran directed one of his teams to the front of the house, the other accompanied him as he

41

approached the two men. They were dressed not unlike a typical local, in blue jeans and tee shirt, and greeted our approach nonchalantly. More physical detail came into focus as we drew near. Both men were well built, had close haircuts, and appeared to be in their early to mid-thirties. Approaching closer still, we all noticed that both men were wearing U.S. Army jungle boots, which were typically worn by the Special Forces. Sergeant Beran hailed them. "Hey, how's it goin?"

"Fine. We're just getting in a little fishing."

They didn't speak in the dialect of the mountains.

"You guys live here?"

"Yeh."

"Mind if I see some identification?"

With that, both men jumped up, dropping their fishing poles, and ran into the pond. That seemed like a very dumb thing to do. Maybe they thought that the soldiers wouldn't follow them for fear of getting wet. In the event, the soldiers were after them like a pack of hounds after a rabbit, catching up to them in the middle of the pond. Instead of surrendering, they decided to struggle, which was also a bad idea, given that they were tangling with five young paratroopers who had been hoping for just such excitement.

The fight was over quickly, with both fugitives carried, arms pinned behind their backs, and unceremoniously thrown face down on the ground at the feet of Sergeant Jarrette and me. Both men were clearly rattled at the turn of events and seemed unsure about what was going to happen next. One of them, still lying face down with his arms pinned behind him, turned his head up toward Sergeant Jarrette and said plaintively, "You gotta help me, Sarge, I'm a Sergeant First Class in the U.S. Army."

This was a major coup. We now had two prisoners, both senior NCO's from the Special Forces. We had captured their food and equipment and now, we were capturing key individuals in their organization. This was getting better and better.

I'LL CUT YOU

We continued to experience great success against the Green Berets. We captured another air-drop of food. The locals, not to be outdone, captured one of their own, making off with the rations as well as the parachutes and other air items. This would become an issue later on. One day, while patrolling, we came across some footprints with the telltale lug sole pattern of the Army jungle boot. Following the tracks, we found two Green Berets sleeping. Two more prisoners! In my youthful zeal and lack of experience with army exercises, I began to lose sight of our role as training support for the Special Forces exercise, instead seeing this as a contest that we could win.

But other events were taking place as well. One day we got a call on the radio that the battalion commander, Lieutenant Colonel Joseph Gilmore, was flying up from Fort Bragg to visit us and that he and Captain Ferguson would be coming around to the platoon command post sometime during the day. I hadn't seen or heard from Captain Ferguson since my arrival in the field, nor did it seem that anyone else had either. Rumor had it that he was hanging around in the town of Boone.

This rumor seemed to be born out when Colonel Gilmore and Captain Ferguson arrived at our camp. Visible under Captain Ferguson's fatigue shirt was a blue polo shirt. It appeared that Colonel Gilmore's visit had so caught him by surprise that he barely had time to throw his fatigues on over his civilian clothes. I took the battalion commander on a tour of our little camp and briefed him

43

on what we had been doing operationally. He seemed satisfied, leaving us with words of encouragement.

One night, I got a call on the radio informing me that Doc Durnig and Specialist Patrick McGrath had been in a wreck and that McGrath was injured. Both Durnig, our medic, and McGrath, were from Philadelphia. McGrath was a tough Irish kid with a bulldog face that only a mother could love. It seemed that every time he returned from a weekend pass back to Philly, his hands would be bandaged from having been in a fight. For all that, he was one of those characters who provided the heart and soul of the platoon.

Sergeant Jarrette and I jumped in our vehicle and sped off to the regional hospital in Boone. Upon arrival, we rushed into the Emergency Room, where we found McGrath laid out on an examining table with his head sandwiched by sandbags. There was concern that he had a neck or spinal injury.

"Hey McGrath, how you doin'?"

"Yeh, hey sir. I'm OK."

"What're you doing to yourself? You know, we might have to write you up for damaging government property."

"You'll have to send the bill to Durnig, he was driving."

While we were making small talk, the nurse came in.

"OK, we want to get you in to X-ray and have a look at your neck. Let's see, were going to have to take that necklace off before we can X-ray you."

McGrath was wearing a small cross on a chain around his neck.

"No! Don't take that off! My girlfriend gave that to me. Don't touch it!"

"But sir, we have to take it off in order to do an X-ray."

"No! You better not touch it! That was a gift from my girl."

A volcano was beginning to erupt from this inert body with

its head encased in sandbags. Quick as a flash, McGrath reached into his pocket and came out brandishing a tiny P-38 can-opener. These were the can-openers that came with our C-rations, 1 ½ inches long with a half-inch blade.

"You know what this is? This is a P-38. I'll cut you. I'll cut you bad!"

Now Sergeant Jarrette and I were galvanized. To us this was absurd and amusing, but not so much so for the nurse. Signaling the nurse to back off, soothing words were spoken. "Whoa McGrath. Calm down. No one's going to take your necklace. Relax. Put the P-38 away."

Now the nurse re-entered the fray. "I'm going to give you a shot that will relax you, OK?"

Sergeant Jarrette and I spoke reassuringly. "She's going to give you a shot that will calm you down. Just relax."

McGrath submitted to the shot and soon enough was sleeping like a baby. The X-rays were taken and, fortunately, showed no damage to his neck. He'd be in a collar brace for a few days, but other than that, he was OK.

THAT'S LIEUTENANT WILSON'S CAR

Over the next couple of days, Sergeant Jarrette and I continued to hear amazing stories from our teams as we made the daily rounds. Finally, I couldn't stand it any longer, saying to Sergeant Jarrette,

"This is killing me, I've gotta see this for myself. Let's come out and spend the night."

After some discussion, we decided that the bridge where the mountain girls were showing up offered the most interesting prospects. Little did we know.

Having packed our rucksacks for an overnight stay, we arrived at Sergeant Hosey's bridge in the late afternoon, setting up camp on the stream bank underneath the bridge. As if on cue, the two girls arrived shortly thereafter. They were just as Sergeant Hosey had described them, like characters out of a Lil' Abner comic strip. It actually seemed a little awkward, as if Sergeant Jarrette and I were unannounced visitors at someone's family gathering. Perhaps the girls felt the same, for they didn't stay for very long.

After a dinner of C-rations, we relaxed in the cool of the autumn night. Sergeant Jarrette regaled us with stories of Command Sergeant Major Pearce, who had been his platoon sergeant when he was a private. CSM Pearce was a hard-boiled veteran of WWII, Korea, and Vietnam and was currently the senior NCO in the 82nd Airborne Division. Sergeant Jarrette referred to him as "the man with the upside-down smile" and could do a hilarious imitation of his speech and mannerisms. Other stories were told of Colonel "Smoke", our previous battalion commander and of the legendary Command Sergeant Major Sabalauski, another tough veteran of three wars.

Around nine o'clock, everyone having exhausted their repertoire of war stories, we began to settle down for the night. Specialist Waggoner was walking guard up on the bridge, the rest of us laying against our rucks, enjoying a cigarette before sleep overtook us.

All of a sudden, a car came flying off the bluff on the other side of the stream where the road came down the mountain and turned onto the bridge. It landed nose first on the embankment, tipped over upside down in the stream and burst into flame.

With a collective shout everyone was on their feet and running in the direction of the burning car, the passenger compartment of which was under water. As I got to the water's edge, I could see that it was a red sports car.

"Holy shit, that's Lieutenant Wilson's car!"

Lieutenant Wilson was one of the other platoon leaders in the company.

Sergeant Jarrette and others were shouting and splashing their way to the car, pulling open the passenger side door, which was closest. I was on the bank preparing first aid dressings and yelling at Hartsfeld, my radio operator.

"Get him over here! Hartsfeld, call the company and tell them to get an ambulance out here ASAP!"

While Sergeant Jarrette made his way around to the driver's side, Sergeant Hosey led the first occupant over to me, whom I recognized as Specialist Bancroft, the assistant to the company operations sergeant. His head was split open from one side of his forehead to the other. In spite of the emergency of the moment, I found myself wondering what Lieutenant Wilson was doing out joyriding with a junior enlisted man.

"Come on. Let me take a look at you."

Guiding him back away from the bank, I placed the dressing on his wound, tying it down tight to staunch the flow of blood. I then had him lay down and prop his feet up on someone's pack, covering him with a field jacket.

"Hartsfeld! Did you get hold of the company?"

"Roger that, sir. They've contacted the hospital and there's an ambulance on the way."

"OK, good."

Meanwhile, Sergeant Jarrette was leading Lieutenant Wilson over to the bank. His head was also opened up, even deeper and wider than that of Bancroft. After administering the same first aid to him, I said, "Boy, you two are lucky that we happened to be here."

I didn't think it was an appropriate time to ask him what the hell he was doing out tearing around the mountains with a junior soldier.

The ambulance showed up after a while and packed the two injured men off to the hospital.

Wide-awake now, we resumed some animated chatter. "It's a good fuckin' thing we were here or those two would've drowned."

"That's for damn sure."

"They must have been going pretty fast and missed that turn just before the bridge."

"They're both going to look like Frankenstein with their heads stitched up."

"Yeh, the Frankenstein twins."

The question of why a lieutenant was hanging out with a soldier was politely avoided.

After a while, the banter quieted down and we gradually drifted off to sleep, except for the guard and the radio watch.

THE MIDNIGHT HOOK GOT HIM

Two days later, I was back in the platoon CP when Lieutenant Ron Hendrickson, the company Executive Officer showed up. Ron was a big, serious, straight-laced Texan and also a good friend. We called him Brawny because of his resemblance to the actor in the commercials for Brawny paper towels.

"Brawny! What brings you out here?"

"Captain Ferguson got relieved. I'm in temporary command."

"You gotta be kiddin'. Where is he now?"

"I don't know. He's just gone."

"Sounds like the midnight hook came and got him. What's going to happen now?"

"Major Long, the new battalion XO, is coming up to take charge of the company until we re-deploy."

Relieving Captain Ferguson was the best thing that could have happened to the company, and it was long overdue. And yet, here we were, like children in an alcoholic household, feeling guilty and responsible for his demise. We were also feeling fearful of this unknown major who was coming up to take charge. Like children, our deepest fear was, "What does this mean for me?"

The next day, Major Long arrived and assembled the officers and senior NCO's. Would he blame us? Would he hold us responsible? Would he yell at us and threaten us? As we assembled, I got my first look at him. He was good looking, a bit like Paul Newman, and he had prematurely silver hair. He would later become known as the Silver Fox.

We formally lined up, but he immediately put us at ease. "At ease. Come on in here."

He beckoned us to come close to him where he stood by the front of his jeep, cap off, one foot up on the bumper, his field jacket unzipped and open in the front. He seemed to instinctively know how we were feeling and was taking steps to relieve our anxiety.

"I'm Major Long and I've just joined the battalion as XO. I'm going to take charge of things up here until we get back to Fort Bragg. Look, this isn't your fault..."

That was all we needed to hear. He was talking to us man-to-man, soldier-to-soldier. He wasn't treating us like misbehaving children. Whatever other instructions he gave us were gladly received and complied with. We still had a mission to accomplish and that's what he wanted us to do, and we were happy to do it.

FINALE

After the trauma of Captain Ferguson's relief, life resumed as it had before, with patrols and the daily rounds of my positions. This exercise had been going on for quite a while, so I was pretty sure we were approaching the grand finale, with the Special Forces making their move to seize Boone. I was also pretty sure that this would involve an attack on the big bridge being defended by Staff Sergeant Nathan Carter's squad. I alerted them to be at a high state of readiness, but, beyond that, I chose not to over-supervise by placing myself on the bridge.

In my view, we had punished the Special Forces badly and I was convinced we could beat them. In fact, I was bent on it, so bent on it that I made a stupid mistake.

One day I arrived on the bridge during my rounds and Sergeant Carter informed me that they had just been hit by the Special Forces. I blew my top.

"What do you mean, 'you got hit?'"

"They just hit us, sir."

"They caught you half-stepping, that's what they did. If you hadn't been sitting on your asses, you would have kicked their ass."

This was said in the presence of the whole squad. Sergeant Carter was a professional soldier and knew how to show no emotion while being berated. The soldiers in his squad were not so restrained, especially Specialist Huntley, a Vietnam veteran who had been shot seven times less than a year previous in one of the major battles of that conflict.

"Fuck this. I don't want to be in this platoon no more. I want a transfer."

I immediately sensed that I had made a mistake, but it was too late, the words had already been spoken. A short while later, after I had cooled down, Sergeant Jarrette said to me, "They were going to hit us, no matter what, sir. We're just training aids for them, and the purpose of the exercise is for them to win."

A good platoon sergeant is, among other things, a mentor to his young lieutenant. Sergeant Jarrette's tone was not scolding or condescending. He merely provided me with information and left it up to me what to do with it.

There was in the Army an unofficial adage that an officer should never apologize to his men for any reason. I was still learning the ways of the Army and trying to make sense of them, but there was something about this dictum that didn't seem right. If you make a mistake, shouldn't you be willing to say so? If you have caused damage, shouldn't you try to repair the damage?

I gathered Sergeant Carter's squad. "Look, I apologize for losing my cool a while ago. I know you guys have been working hard and doing your job. I just got too carried away with the exercise and I over-reacted. Anyway, I'm sorry."

I don't know what the wisdom was behind never apologizing, but in this case I saw no discernable negative effects for having done so.

Now that the exercise was pretty much over, there remained the issue of the airdrop that had been captured by the locals. The cargo had consisted entirely of C-rations, which the powers had no interest in retrieving. The parachutes and rigging were another matter. I was told we would not leave without them.

How to find a needle in a haystack? Ask the person who might know where the needle is. I approached the farmer on whose land we had set up our camp, the same farmer who had produced a jar of "good moonshine" so as to instruct us on the good from the bad.

"During the exercise, one of the airdrops got captured by the locals. We need to get it back."

"Well, I wouldn't be knowing about that."

Something evasive in his tone of voice suggested he did know about it, or he knew how to find out.

"Well, here's the deal. We don't care about the C-rations. Whoever got them can have them. But the air items, the parachutes and rigging, we have to get back. We can't leave until we find them and if we don't find them soon, there will be more people coming up here to help look for them."

The last thing the locals wanted was a bunch of people snooping around the mountains where they had their moonshine stills.

"I'll see what I can find out."

A day or so later, he came back with instructions to meet some people at a given time on a dirt road deep in the woods. At the appointed time and place, we met some of the local gentlemen, who produced from their pickup truck the precious air items. The mafia had nothing on these guys.

On the last day before re-deployment, we had a chopper blast—a parachute jump from a helicopter. Jumping from a plane involved several hours of discomfort and airsickness for about one minute of excitement. A helicopter jump was more like an amusement park event. The helicopter would land right next to you, you'd get on, the helicopter would fly up to the appropriate altitude and you'd jump. Fifteen minutes from start to finish. Lots of fun.

Thus ended one of my more colorful experiences in the Army. A month later, I was ordered to Vietnam.

Author and SFC John Jarrette Jr., Boone NC

6. GI Humor

One of the unique features of military service is it takes in-dividuals from completely different walks of life and throws them together, creating the possibility for friendships that likely would not have occurred otherwise. Such was the case with Ruben Bugge and Charles Warner.

In physical appearance, Bugge and Warner were as different as day and night. Bugge had bright red hair, a fair complexion, and freckles. Warner, on the other hand, had dark, almost black hair, a large moustache, and heavy-browed, dark eyes. He could have easily played the role of the villain in an old time western movie.

Ruben Bugge, known to his friends as Ben, was an easy-going kid from the middle class suburban city of Novato, California. In 1971, he found himself bored and without direction, so he enlisted in the Army and volunteered for service in Vietnam. His wish was granted and, six months later, he found himself as a grenadier in a rifle platoon in Vietnam.

In this platoon, Ruben met Charles Warner. They shared an easy-going nature, but beyond that their backgrounds were quite different. Warner was a wild, cowboy type from somewhere in Colorado. He had been in the Army for a few years and had risen to the rank of Staff Sergeant but had been busted down to buck private for some unknown offense. He held his cards close to his chest and never spoke of his background. This reticence, combined with the little that was known of him, added to his mystique. He seemed to relish being in Vietnam. On one arm he had the word 'War' tattooed and he wore a medallion that said 'War', strapped to his wrist with a wide leather band.

Behind his easy going façade there lurked a mischievous mind always on the lookout for a prank.

In spite of their differences, Bugge and Warner quickly became buddies. Besides being squad mates watching each other's back, they ate together, hung out together, and shared the same fighting position when in the defense.

In March 1972, Bugge and Warner's company was brought in from combat operations in the mountains to the west and assigned to man static defensive positions on what was known as "The Ridgeline", which was the first line of defense for the city of Da Nang. Unlike normal combat operations, which were characterized by constant patrolling and movement, there was little to do on The Ridgeline other than fill sandbags and stare out from their bunkers at a landscape of rice paddies and rolling hills, devoid of human activity. Even though subject to attack, and therefore potentially dangerous, the lack of activity produced acute boredom and, as the old adage goes, "idle hands are the devil's workshop."

In order to relieve the boredom, individual soldiers were rotated to the rear to get a shower and change clothes and maybe even go to the nearby air base to partake of the good life enjoyed by U.S. Air Force personnel. Another way to provide diversion was to send soldiers on details to the rear to pick up supplies.

One very hot day, Bugge returned from a supply run to the rear. He was tired, hot and sweaty. As he approached the bunker that he and Warner shared, he called out, "Hey, who wants a cold soda?"

He got no response and, as he got closer, he could see Warner and Sergeant David "Rock" Mixon sitting facing each other, almost knee to knee, just inside the bunker entrance. Closer still, Bugge picked up the sounds and vibes of a heated argument between the two. This struck him as odd because he had never seen Warner get angry at anyone.

In spite of the awkwardness of the situation, Bugge wanted to get into the bunker and away from the sun and heat. In order to do so, he had to slide between the knees of the two belligerents and take a seat near the back wall of the bunker.

Sergeant Mixon, his face flushed with anger, and his mouth twisted in a snarl, said, "Warner, I'm gonna kick your goat smellin' ass!"

"I'm ready whenever you are, Mixon."

With that, Sergeant Mixon pulled a .45 out of his waistband, pointing it at the ground but holding it in a menacing manner.

Alarmed, Bugge said, "Dudes, calm down! What the hell's going on?"

Ignoring Bugge as if he wasn't even there, Warner pulled out a hand grenade, staring back at Sergeant Mixon with equal menace.

"Chuck! Whoa. Put that away, man. This shit's getting way out of hand. Calm down."

No sooner were the words out of Bugge's mouth than Warner pulled the pin on the grenade. The spoon came flying off and he let the grenade drop between him and Sergeant Mixon.

As happens in moments of extreme crisis, time slowed as a myriad of thoughts raced through Bugge's mind.

'There's four and a half seconds until it explodes.'
'Surely one of them is going to throw it out of the bunker.'
'Surely they're not going to stay here and get blown up.'
'I'm gettin' the fuck outta here.'

Yelling, "Are you crazy!? What're you doin'?" Bugge dove over their knees and through the entrance, landed on his belly and low crawled through the blistering hot sand as fast as his knees and elbows would carry him, all the time waiting for the explosion that would kill his two buddies and maybe him. Just at the moment he expected the detonation, he heard, "Ha, ha, ha, ha." Raucous laughter blasting from inside the bunker.

Picking himself up and dusting off his arms which were caked with sand and skinned, Bugge turned back to find Warner and Sergeant Mixon roaring with laughter.

"Oh man, you should have seen yourself. All we could see was assholes and elbows when you di-di'd the bunker. What a trip. You better check your drawers."

Sputtering with rage, all Bugge could manage to say was, "You fuckers, you mother fuckers."

Calming down, Bugge suddenly became fascinated with the mystery of why the grenade didn't explode. "What happened? How come it didn't go off?"

Chuckling, Warner replied, "I just broke off the blasting cap. There was nothing to detonate the explosive."

And so ended this little practical joke, a bit of GI humor, the story of which never made it out of this trio of pals until almost fifty years later when Bugge shared it with his former platoon leader.

Boys and their toys, you just never know what's going to happen.

L-R: PFC McCartney, SP4 Ruben Bugge, PVT
Chuck Warner, SGT David Mixon

7. There Ain't Been Nobody Up There

The recent fiftieth anniversary of the moon landing got me to thinking about it and events in my life associated with it. First of all, I remember exactly where I was when it happened on July 20, 1969. I was in basic training at Fort Lewis, Washington and we were on a twenty-mile march. One of the drill sergeants set up a small battery operated television at the side of the trail so that we could see the landing as we passed. Marching by, I marveled at the comparison between the two events. The Roman Legions traveled twenty miles a day when on the march and here we were doing the same thing. In fact, for millennia soldiers have moved long distances by foot, and here were men walking around on the moon.

Thinking of the moon landing also makes me think of Frank. Frank was my next-door-neighbor in Fayetteville, North Carolina, after I returned from Vietnam in 1972. He was a World War Two veteran, probably in his mid-fifties. He worked for the gas and electric company, presumably as a lineman, since his tool truck was usually parked at the house.

One night, I went out to the front yard and encountered Frank at his tool truck. The night was clear, lit with stars. A brand new moon hung vertically in the night sky.

"Hey, Frank."

"Hey."

Looking up at the new moon, Frank continued. "Looka there. They ain't been nobody on the moon. Look at it. You'd fall off."

He was, I assumed, referring to the slender crescent shape of the new moon.

"I'll tell you what they been doin'. They been down there in Arizonee and New Mexico jumpin' around in the desert and takin' pictures of it. They ain't been nobody up there."

Well, there you had it. The voice of authority. What could I say?

Early on a cold Sunday morning in February 1973, I was in the back yard digging when I heard Frank call out to me. "Hey, come on over here. I want to show you something."

Welcoming an excuse to drop my shovel, I walked next-door, curious about what it was Frank wanted me to see. Entering the kitchen and greeting Frank's wife, I was enveloped by warmth and the aroma of fried bacon and fresh brewed coffee.

While I stood there wondering what I was to be shown, Frank reached into the kitchen cupboard, pulled out a bottle of whiskey, and proceeded to fill a water tumbler full of the amber liquid.

"Can I pour you some?"

"Uh, no thanks, I haven't had breakfast yet."

"Suit yourself. Wait here just a minute."

A few moments later, Frank reappeared carrying a rifle. At a glance I identified it as a World War Two era German Mauser, the standard infantry weapon of the Nazi army. Handing it to me, I now noticed the stock, which was beautifully carved into a forest scene, complete with deer, pheasant and evergreen trees. It was just like the intricately carved stocks that one might see on expensive shotguns or hunting rifles in Abercrombie and Fitch, back when it was a millionaire's toy store.

"Wow, Frank, this is beautiful. Where in the world did you get it?"

"A German POW carved it for me during the war. He did it for a carton of cigarettes."

"A carton of cigarettes!!? Dang! What a prized souvenir. Speaking of POW's, what do you think of the release of the POW's from North Vietnam?"

"They ain't been no prisoners."

Gob smacked, I said, "What do you mean?"

"My unit liberated some of them concentration camps in the war. I've seen prisoners. These ain't been no prisoners."

He didn't elaborate, but I assumed he meant that the returning POW's didn't resemble the living skeletons that he had encountered in the Nazi concentration camps.

As with our discussion of the alleged moon landing, there was no sense arguing with the voice of authority. After a little more small talk, I made my move to head home. Handing the rifle back to Frank, I said, "Well, I'd better get back to my chores. Thanks for showing me this. It really is a beauty. Enjoy the rest of the day."

Perhaps there is a lesson here for those well-informed, urbane citizens who can't understand the way other people think, act, and vote. For many people, if they haven't seen it with their own eyes, it didn't happen. Frank was no Holocaust denier. He'd seen it. Landing on the moon? Well, that was another story.

8. Baptism

Catholics have the sacraments of Baptism, First Commu-nion, and Confirmation for their rites of initiation into the faith. Jews have circumcision, Bar/Bat Mitzvah, and confirmation for theirs. Many religions and institutions have initiation rites, but none of them hold a candle to the initiation into the U.S. Army airborne — the Prop Blast.

On 25 October 1972, my battalion, the 2d Battalion, 508th Airborne Infantry published PROP BLAST ORD 1-73, which announced that a Prop Blast for those not previously inducted into the "Honored Circle of Prop Blasted Jumpers" would be held six days hence, on 1 November.

Included in the order were specific instructions as to group assignments, uniform to be worn, reporting times, etc. Additionally, there was a long list of technical questions that we were required to answer correctly, such as:

- How many gores are there in the canopy assembly of the T-10 reserve parachute?
- What is the tensile strength of one suspension line?
- What is the width and length of the M-1950 weapons case?

And many, many more of similar technicality.

Finally, we were to commit to memory all eight lengthy verses to the paratrooper song "Blood on the Risers," and the 82d Airborne Division song, "The All American Soldier."

I was assigned to Chalk 1, along with my best friend, 1LT Jeff Guild, the battalion XO, Major Gerry Griffin, and two other lieutenants.

So, at 1500 hours on 1 November 1972, all five chalks, each made up of five officers, assembled behind the battalion headquarters, as ordered. First, we were inspected to make sure we were in the specified uniform, which consisted of fatigues and boots, full web gear with combat pack, steel helmet, dog tags, and shot record.

Inspection and manifest call complete, we were off and running—literally.

"Group, Atten-shun"

"Left face."

"Forward, march."

"Double-time, march."

As we ran, we sang the traditional running songs, known as 'Jodies'.

C-130 rollin' down the strip,
Airborne daddy gonna take a little trip.
Stand up, hook up, shuffle to the door,
Jump right out and count to four.

Our destination was Trainer's Tavern, which was an annex to the Officer's Club located in the "old division area", a cantonment area made up of WWII era wooden barracks and ancillary structures.

After running for about 30 minutes, we arrived at Trainer's Tavern, whereupon we began a two-hour period of physical training (PT) and drinking in the large grassy area behind the tavern. PT consisted of calisthenics and wind sprints interspersed with large slugs of vodka.

Captain Larry Ellis, the battalion S-3 for whom Jeff Guild and I worked, was also our PT instructor. Sitting with our backs against the outer wall of the tavern in between exercises, Captain Ellis came along with a bottle of vodka. "Open up."

Opening our mouths, he would tip the bottle up and pour in the vodka. "Goddamn, Jeff, is he trying to kill us?"

"You know he's a teetotaler. He probably has no idea how potent this shit is."

"Well, I do. Next time he comes by, I'm going to hold it in my mouth and spit it out when he isn't looking. I don't want to get shit-faced."

And that's what we did. We certainly weren't sober, but we had no desire to get legless drunk.

The physical hazing and drinking continued for about two hours until, at 1800 hours, we were gathered to hear the history of the Prop Blast, how it began in1940 at Fort Benning, Georgia with thirteen of the original paratroop officers, and how it has continued ever since, in peacetime and in war.

At 1830, we were instructed to draw parachutes. "Line up in chalk order. Starting with Chalk 1, report to the back of the Gamma Goat over there and draw one T-10 main parachute and one T-10 Reserve. 'Chute up. Once everyone has donned their chutes, there will be an in-ranks jumpmaster inspection prior to boarding the aircraft for the jump. Is that clear?"

"AIRBORNE," we yelled back in unison.

At 1900, the inspection complete, the OIC, Captain Dave Mc-Millin, gave the command, "Chalk 1, follow me."

Led by Major Griffin, our chalk leader, we entered Trainer's Tavern. There, inside the building, was a life-size mockup of a C-130 fuselage, with the jump door close by. Beyond, sitting at a long table, was the board of officers, all decked out in Class-A uniforms. The Honorary President was our commanding general, Major General Frederick "Fritz" Kroesen; the President, 1st Brigade Commander Colonel Edward Partain, my CO of 2-508, Lt Col Joe Gilmore; the CO of 1-504, Lt Col Mike Lally; the CO of 2-504, Lt Col Rocco Ventrella and, finally, my future CO, Lt Col Pat Leighton.

Captain Mc Millin gave the following instructions: "When the jumpmaster tells you to 'Stand in the door', you will remain there until given the command 'Go'. Once you have exited the aircraft, you will perform the following functions: one, you will check your body position and count; two, you will check your canopy; three, you will keep a sharp lookout during descent; four, the DZSO, Captain Ellis, will call out what kind of landing you are to make, tree, water, etc. and you will prepare for that type of landing and, finally, you will land, performing a proper PLF (Parachute Landing Fall). Is that clear?"

"CLEAR, SIR, AIRBORNE!" we answered back.

"Board the aircraft."

My manifest number was 5, so I was the last person in our chalk, which meant I would be the last of our little group to jump.

Once we were seated on the jump seats, the jumpmaster began the jump commands.

"GET READY."

"STAND UP."

"HOOK UP" (simulated).

"CHECK EQUIPMENT."

Each of us checked our own equipment and the parachute pack of the person to our front.

"SOUND OFF FOR EQUIPMENT CHECK."

From rear to front, slapping the butt of the person in front, each of us yelled "OK" until it got to the lead person who pointed at the jumpmaster and yelled, "ALL OK".

"STAND IN THE DOOR."

Major Griffin pivoted to the right and took up the door position, slightly crouched, left foot leading, hands positioned outside the fuselage, ready to launch himself when the command "GO" was given.

After what seemed like a full minute of standing in the door, the command "GO" was given and Major Griffin disappeared from view. I could hear him counting, "One thousand, two thousand ..." as well as lots of yelling. And so it went for the other members of my chalk. Finally, it was my turn.

"STAND IN THE DOOR."

I shuffled forward in the prescribed manner, pivoted when I reached the door, and slapped my hands on the outer surface of the door, and ...

BZZZZZZZZZZZZTTTT ...

Holy shit! I was getting an electric shock! Out of the corner of my eye I could see someone madly cranking the handle of a TA-312 field telephone which was wired to the door. No wonder they were keeping everyone standing in the door for so long! This explained the strict admonition we received at the beginning not to jump until given the command, "GO".

After what seemed like an eternity, I heard the command "GO!" and launched myself into space, automatically performing the points of performance: tucked into a tight exit position and counting aloud, "One thousand, two thousand, three thousand, four, thousand," then spreading my arms wide and looking up to check my canopy. Then I heard the commanding voice of Captain Ellis, "WATER LANDING."

I jettisoned my helmet and then began frantically unhooking my reserve parachute and preparing to release my parachute harness so that I could slip out of it once I hit the water. As I was doing this, I was getting drenched by buckets of water being thrown on me. Then I did a PLF and leaped to my feet. Captain Ellis disdainfully said, "Report back to the marshalling area."

It was immediately clear that no one was going to pass on the first try, or maybe even the second or third. So I went back outside and joined all the other members of my chalk where we waited for the other four chalks to have their turn.

After a couple more jumps which resulted in banishment back to the marshalling area, I was called forth yet again. It was now around 2100 hrs.

The jump commands were all repeated and I received the command to stand in the door. Enduring the electric shock was slightly easier to endure given that I expected it, but it still wasn't fun. Then the command, "GO!"

I leaped out the door, assumed the proper position and went through the standard points of performance. This time, Captain Ellis yelled out, "TREE LANDING."

As I was pulling myself into a good, tight body position, chin to chest, I received a whack, then another and another. I was being flailed with a tree branch as I went through the tree landing procedure. Going on the assumption that I was caught up in a tree, I did not execute a PLF, but continued to bounce around on my toes like a pogo stick, expecting at any moment to be ordered back to the marshalling area.

Then I heard, "REPORT TO THE PRESIDENT OF THE BOARD!"

I double-timed over to the table where the board was seated, came to attention, saluted, and reported to the president in the prescribed manner.

"Sir, Blastee Byrne Sherwood, 566-23-4567, respectfully reports to the President of the Board and very meekly and humbly requests consideration for acceptance into the traditional order of Prop Blasted Jumpers."

Colonel Partain returned my salute but left me standing at the position of attention.

"What is the significance of September 17th?"

"Sir, 17 September is the anniversary of Operation Market-Garden, when U.S. and British airborne forces jumped into Holland."

"That is correct. Colonel Gilmore, do you have any questions?"

"Yes, sir. Blastee Sherwood, sing the second verse and chorus of 'Blood on the Risers.'"

Lt Col Gilmore was my CO and had given me an easy one. We all knew this airborne anthem, sung to the tune of the 'Battle Hymn of the Republic.'

"Yes, sir."

Is everybody happy cried the sergeant looking up.
Our hero feebly answered yes and then they stood him up.
He jumped into the icy blast, his static line unhooked,
And he ain't gonna jump no more.
Gory, gory, what a helluva way to die.
Gory, gory what a helluva way to die.
Gory, gory what a helluva way to die.
And he ain't gonna jump no more."

"Well done, Blastee Sherwood. Colonel Ventrella?"

"Blastee, what is the length of the static line?"

This question, too, was relatively easy. "Sir, the length of the static line is fifteen feet."

"Correct. Colonel Lally?"

"Blastee, what is the minimum altitude for airdrop of personnel on tactical training jumps using the T-10 parachute?"

I couldn't believe it. Of all the highly technical information we had to memorize, I was getting some really easy questions.

"Sir, the minimum altitude for airdrop of personnel on tactical training jumps using the T-10 parachute is 1,000 feet."

"Very good, Blastee. Colonel Leighton?"

Lt Col Pat Leighton, who was soon to replace Lt Col Gilmore as CO of my battalion, was a severe looking man who seemed to take this all very seriously. Looking at me hard, he said, "Blastee, what is the tensile strength of one suspension line?"

My intoxicated brain started frantically reviewing all the questions and answers, the majority of which were numbers: units of weight, distance, length. Was it, 24, 1000, 95, 375, 15, 18 or 30? Should I guess? If I get it wrong, will they send me back out to the marshalling area to start all over again? I opted for the standard Army response when one didn't know.

"Sir, I don't know, but I'll find out."

Looking at me like a parent whose child has just come home from school with a bad report card, Lt Col Leighton handed it back to Colonel Partain.

"Blastee, sing the final verse to 'Blood on the Risers."

"Yes, sir."

There was blood upon the risers,
there were brains upon the chute.
Intestines were a dangling from his paratrooper suit.

He was a mess, they picked him up
and poured him from his boots,
and he ain't gonna jump no more.

What now? Back to the marshalling area?
"Blastee Sherwood, report to the Keeper of the Crock."

Each airborne regiment had its own Prop Blast mug, or crock, which was like the holy grail for that regiment. Each had a name, like "Miley Mug", "Gavin Goblet", "Sink Grail" or "Zipper Dipper". Ours was called the "Lindquist Liberator," named after Col. Roy Lindquist, the WWII commander of the 508th PIR. In appearance, it looked like a large silver trophy cup. The Prop Blast mixture was a combination of vodka and champagne. The vodka symbolized the blast one received when exiting the aircraft on a jump and the sparkle of champagne symbolizing the youth and zest of the airborne. As I approached the Keeper of the Crock, I was fully cognizant that I would be drinking from a container that had been drunk from by every officer in the 508th PIR since WWII. Awe inspiring to put it mildly.

I reported to Captain John Everson, the Keeper of the Crock. He gave the following instructions: "Blastee, I will hand you the crock and, on the command 'GO', you will sound off with your name and drink the crock empty by the count of 'Four thousand'. As soon as you drain the contents, you will hold the empty crock over your head to show that it is empty and shout the first word that comes to your head. Clear?"

"Clear, sir, Airborne."

Handing me the Lindquist Liberator, he shouted 'GO'.

I sounded off with my name and started chugging the contents for all I was worth. There was a tiny crack at the base of the lip of the cup, leaking some onto my shirt front, but the bulk went down my gullet. As I drank, I could hear the board loudly chanting in unison, "ONE THOUSAND, TWO THOUSAND, THREE THOUSAND, FOUR THOUSAND."

I emptied the Liberator just as the board hit 'Four Thousand', I held it over my head and yelled, "AIRBORNE!"

"Well done, Blastee. Report to the Recorder and sign the book. Once you have done that, report back to the marshalling area, turn in your parachute, and return here to be a spectator.

"Clear, sir, Airborne!"

And so, I was fully baptized into the brotherhood of paratroopers. I had undergone the hazing and drunk the same concoction from the same container as all those heroes who had gone before. It was at once a humbling and exhilarating feeling.

A little over a year later, I was on the Prop Blast Committee as a Safety Officer, PT Instructor, and Tree Landing OIC as we baptized a new crop of officers into the Airborne brotherhood.

9. Big Clark

What are you most proud of in your professional career?
For a professional soldier, you might imagine it to be the rank achieved, highest organization commanded, or perhaps medals awarded. For many, however, it is something seemingly much smaller. For me, one of the achievements of which I am most proud is Big Clark.

After my return from Vietnam, I was again assigned to the 82d Airborne Division at Ft. Bragg, North Carolina, from whence I had departed a year earlier. I managed to get assigned down to the same battalion I had been in before. This was very therapeutic for me. Vietnam left me feeling at loose ends, and it was comforting to go back to the 2d Battalion, 508th Parachute Infantry where I knew everyone and they knew me. It felt like home.

Initially, I was assigned to the battalion staff in the Operations and Training section, along with my old friend Jeff Guild. But, after ten months of staff work, I felt a deep longing to get back to a rifle company to be around soldiers. The Battalion Executive Officer, Major Gerry Griffin, was good enough to heed my request and I was assigned as the company executive officer of Company C, under the able leadership of Captain Pete Tolley and First Sergeant Willie Roach.

Very quickly after my assignment to Charlie Company, I noticed that the same person, Staff Sergeant Clark, was CQ (Charge of Quarters) every night. Normally, the duty of CQ, the person in charge of the company during after duty hours, was rotated among the Staff Sergeants and Sergeants, according to an impartially run duty roster. The idea of having one person perform this unpleasant duty was a curious deviation from normal procedure. I decided to find out by asking the First Sergeant.

"Top, how come Sergeant Clark is on CQ every night?"

"He's got a medical profile which prevents him from physical activity and field duty. So, we just keep Big Clark on CQ. He doesn't mind, and it spares the line NCO's from having to do it."

"That makes sense. Thanks, Top."

The logic was unassailable, and I accepted it at face value.

Since Sergeant Clark came on duty as we were going off duty, it took me a while to get to know him. Appearance wise, there was not much to recommend him. He was an older man, probably close to forty, and only a Staff Sergeant when he should have been at least a Sergeant First Class. Instead of the heavily starched uniforms that we wore, his were washed in the washing machine and fluff dried. Instead of wearing paratrooper boots spit polished to a mirror finish, he wore GI boots that were brush shined.

However, behind this unprofessional façade, there was something appealing about Big Clark, (so named because he was indeed big, over six feet tall and thickly built). He was always polite and friendly, exuding a folksy wisdom and quiet dignity that one often encounters in older Black men from the South. A comparison to some of the characters played by Morgan Freeman wouldn't be far from the mark. Even more intriguing, I began to hear rumors about him from the troops, rumors laced with awe and respect.

"Big Clark this and Big Clark that."

"So and so came in drunk last night raising hell, and Big Clark knocked him out and put him to bed."

I was drawn to find out more about this anomalous character. How better to learn more than from the man himself? Hanging around after duty hours, I began my inquiry. "Hey Sergeant Clark, how's it going?"

"Can't complain, sir, and how about you?"

"Outstanding. Hey, I've been wondering, how come you're on permanent CQ?"

"I've got a medical profile that prevents me from performing in my MOS (Military Occupational Specialty)."

"Really? What's your MOS?"

"I'm Eleven Charlie, mortars."

"So what is it that you can't do?"

"The bones in my feet have collapsed. I can't run or march."

"Surely they can fix that?"

"I don't know. No one ever said anything about fixing them. They just put me on a permanent profile. I'm non-deployable and non-promotable, so, here I sit."

"That doesn't sound right to me. How long have you been in?"

"Oh, twenty one years now."

"Twenty-one years! Were you in Korea?"

"Yes, sir. I was just a young buck then."

"What unit were you in?"

"I was in the 187th Airborne RCT and the 45th Infantry Division."

"No kiddin'? What was that like? I bet there weren't too many black guys in the 187th then."

"Heh, heh. No, sir. We were like specks of pepper on white rice. But it was OK."

"Who were you with in 'Nam?"

"I was with the First Cav."

All of this information would normally have been apparent by looking at a person's uniform, but Sergeant Clark wore on his uniform only those insignia required by regulation — his rank and unit of assignment. Here was the only soldier in the company that had fought in both Korea and Vietnam, and no one even knew it.

Over the weeks and months of casual conversation, I was able to get to know Sergeant Clark and to like and respect him. In order to learn more about him, I pulled his file and discovered that he was the most highly decorated soldier in the company, having earned the Silver Star in Vietnam for braving hostile fire to save a wounded comrade. Although he was not the victim of malice on anyone's part, it seemed to me that he had been brushed to the side for the sake of convenience. Due to his dignity, sense of resignation, or perhaps both, Sergeant Clark raised no protest to his situation. I thought he deserved better.

"I think it's messed up that they won't fix your feet, and I'll bet you it can be done. Would it be OK with you if I look into it?"

"Well sir, you know the old saying, 'Does a bear shit in the woods?'"

With that green light from Sergeant Clark, I dove into my latest extra-curricular project. With a copy of his medical profile and its detailed diagnosis, I called the podiatry clinic and inquired of the Medical Corps captain there why nothing was being done to correct Sergeant Clark's feet.

"There is an operation that can be performed to correct this condition."

"Well, then, why hasn't it been performed on Sergeant Clark?"

"The chief of Podiatry doesn't like to perform them."

Amazed and angry at this cavalier attitude toward the medical treatment that Sergeant Clark needed, I exploded. Technically, I became insubordinate to a senior ranking officer, but I was fairly sure I could intimidate a Medical Corps captain, and I was fully prepared to make good on my threat.

"Whaddayou mean 'he doesn't like to perform them'? Lemme tell you something, I don't think General Kroesen or General Seitz will be too happy if they find out that one of their paratroopers is not being given the medical care he's entitled to."

"OK then, OK. Go ahead and make an appointment for Sergeant Clark and we'll see what can be done."

And so began the process of restoring Staff Sergeant Clark to his rightful place in the Army. I wasn't personally on hand for his appointments, but I checked in with him to make sure things were progressing and not getting bogged down in bureaucracy.

It turned out that he didn't need to have surgery to correct his feet, that orthotic inserts in his boots would do the job. Freed from the limitations of his profile and back on duty now as platoon sergeant of the mortar platoon, Sergeant Clark showed up one day in a brand new uniform, heavily starched with all insignia sewn on, and with highly spit-shined jump boots on his feet. He looked every bit the highly competent non-commissioned officer that he was. Also freed from the block to his promotion, he was soon promoted to Sergeant First Class.

After eighteen months as executive officer of Charlie Company, I was promoted to Captain and assigned back to battalion staff for the few short months remaining before my reassignment to the Infantry Officer's Advance Course at Fort Benning, Georgia.

Shortly before my departure, I remember seeing all the units running down Ardennes Street doing their morning PT run. As Chargin' Charlie Company passed by, there was Sergeant First Class Clark, Big Clark, running alongside his platoon. I felt a warm glow inside.

10. Deep Furrow

The 82d Airborne Division had nine infantry battalions, with far fewer than nine off-post deployments in any given year. To go on any deployment was a major plus, and to go on an out-of-country deployment was to hit the jackpot.

In the summer of 1973, I was still the Executive Officer of C Company, 2-808 Inf (Abn) when we got word that we had been selected to deploy to Turkey in September as part of a NATO exercise called Deep Furrow. The exercise was meant as a show of force to the Soviet Union by the combined forces of the US, UK, Turkey, and Greece. The scheme was for the airborne forces of the US, UK, and Turkey to jump into Turkish Thrace, near the border with the Soviet satellite state of Bulgaria, and then link up with Turkish ground forces.

Simultaneously but unrelated, the decision was made at Department of the Army to replace standard army headgear for airborne units with the maroon beret. Everyone was ecstatic, except for some of the grizzled old NCO's, who were quite happy with the way things had always been. However, there were no maroon berets on hand in the army inventory. To outfit the entire division would require the production of 15,000 berets — a tall order.

Our battalion commander, Lieutenant Colonel James P. "Pat"

Leighton, was determined that we would deploy to Turkey wearing the maroon beret, making us the first battalion in the entire army to sport this new, exotic headgear. Of course, everyone in the battalion enthusiastically supported this idea, (except for those grizzled NCO's).

Part of the preparation for the deployment was intensive briefings on the history, geography, customs, culture, and religion of Turkey. Special emphasis was given to "dos and don'ts". In typical army fashion, scare tactics were used whenever possible. Death was the penalty for being caught with drugs and we were warned that to look directly at a Muslim woman was to risk one's life.

"Man, if they was tryin' to scare the shit out of us, they did a pretty good job."

"Scared the shit out of me. I don't want to get my fuckin' throat slit just because I accidentally looked at a woman."

As the deployment date of 18 September approached, there was still no sign of the coveted berets, and we began to despair of receiving them on time. Then, about one week before the exercise, the berets arrived.

Distribution was rapid. The officers were well aware of how the fashion-conscious paratrooper wears a beret, having carefully observed and enquired about the techniques employed by our European brothers in arms. We speedily boiled them and shaved the fuzz off so that they molded to our heads, hoping that we would appear to have worn them all our lives. The older NCOs tended to just plop them on top of their heads.

"Did you see SFC Stevens? He looks like he's got a big, maroon pancake sitting on top of his head."

"Yeah, he looks like Chef Boyardee."

"Complete with the belly."

Our brigade commander was Colonel Guy S. "Sandy" Meloy.

A recipient of the Distinguished Service Cross, he was a soldier's soldier, respected by all. Unlike many of his rank, he was down to earth and easy to be around. He had once been where we were and had never forgotten. A good leader anticipates what can go wrong and then takes steps to mitigate the threat.

Accordingly, Colonel Meloy directed that no soldier would be allowed to go on pass in Turkey without being issued a condom, with instructions to use it should the occasion arise. As the Executive Officer and second in command of the company, responsibility for such seemingly mundane matters fell to me. Some of my colleagues might have thought it a real hoot to purchase that many prophylactics, but I found it particularly embarrassing.

Duty is duty though, so off to the Post Exchange (PX) I went to purchase a gross of prophylactics. In the aisle where these items were located, all that was there was one open box with individually packaged condoms—not nearly the quantity I needed. I kept looking, as if more might magically appear, but none did. Of course, there would be more in the stock room, but I would have to ask for assistance, adding to my embarrassment. I found the manager's office located up a flight of steps and overlooking the main floor of the PX. In response to my knock on the door, a woman answered. In my 23-year-old prudish mind, this was not a matter to discuss with a woman.

"Can I help you?"

"I need to talk to the manager."

"What do you need to see him for?"

"I just need to talk to him."

Somewhat exasperated, she turned away to fetch the manager from the back of the office. He appeared, looking annoyed at having been interrupted from whatever important activity he was engaged in. "What can I do for you?"

"I, I need a gross of rubbers."

"What!?"

"I said, I need a gross of rubbers. There's not that many on the aisle."

Turning, he called out to the same woman who had greeted me at the door.

"Gladys, get this man a gross of condoms from the stock room, will you?"

'*Well shit*', I thought, '*why not get on the intercom and broadcast it to the whole store? 'Lieutenant Sherwood needs a gross of condoms.'*

Gladys dutifully went to the stock room and came back with a box containing a gross. To my relief, she hadn't asked me what brand or special features I wanted, and I was happy to take whatever she provided, if it was in the correct quantity. Hoping no one would notice, I proceeded to the checkout stand with my precious cargo tucked unseen under my arm. One more obstacle and I was out of there, mission accomplished.

I slipped my single purchase onto the conveyor belt, hoping that this last hurdle would be painless. As the package arrived in front of the checkout clerk, she started to cackle. "Ohhh, hah, hah, hah. Whoa! You the man, you the man."

Blushing mightily and grinning foolishly, I paid up and exited as quickly as possible, glad the ordeal was over. Making a combat air assault in Vietnam was easier.

Finally, the big day came, and we boarded C-141's for the long trip to Turkey. For a trip of this duration, the Air Force put airline seats in the aircraft, facing backwards, which allowed for some degree of comfort. However, unlike civilian airliners, the floors of military aircraft have little, or no insulation and our feet were freezing the whole time. We landed in Rota, Spain but had to remain on the aircraft while refueling took place.

Eventually, we reached Incirlik, Turkey, our Forward Staging Base (FSB).

Incirlik is in southern Turkey near the Mediterranean Sea and the border with Syria. It was desolation personified: flat, treeless terrain all the way to a barren, treeless mountain range almost a hundred miles away. Exceedingly hot and dry, many soldiers got nose bleeds due to their sinuses drying up. When we didn't have any duties to perform, my good friend Jeff Guild, XO of B Co., and I would go to a movie in the air conditioned theater or go sit in the field showers just to try and cool down.

Located with us at Incirlik was a battalion of British paratroopers. The Turkish airborne unit would deploy from their own base somewhere else in Turkey—presumably with a much more hospitable climate. There was much bon homie between us and the Brits—and much horse trading of equipment. The Brits still wore brogans and puttees, and they coveted our high-top boots. We coveted their big, colorful stable belts and jump smocks. I traded a pair of GI combat boots for one of the prized smocks, which were the same design as worn in WWII.

Over and above the operational briefings on the Drop Zone (DZ), assembly areas (AA), and tactical maneuver after the jump, commanders faced the challenge of how to pass the time in this dreadful climate and confined space.

One way was boxing. Matches were organized within companies, between companies and, ultimately, between our battalion and the Brits.

Our champion was a soldier from B Company named Shanklin. The Brits put up a tough looking soldier who looked like he'd been through the mill a few times.

"Are you a boxer, then?" he asked Shanklin.

This was a very subjective question that could be interpreted many ways. Was he a professional boxer, an amateur boxer, or no boxer at all? Shanklin apparently chose to interpret the question as being was he a professional boxer.

"No," he answered.

The fact was that Shanklin had done some Golden Gloves boxing and was no stranger to the ring. Immediately after the bell, there was a flurry of jabs and hooks, and the Brit was down with a broken nose. The Brits seemed to take hardship and injury much more in their stride than Americans and there were no hard feelings about the outcome.

Another way was through Physical Training (PT). Colonel Leighton, apparently wanting to show off his battalion in front of the Brits, ordered a battalion PT run to be conducted after breakfast the following morning. Colonel Leighton was an experienced soldier and should have known better than to organize a run after breakfast, especially in this heat. Normally in the Army, PT was done early in the morning, followed by breakfast.

The following morning, the mess hall served a typical, heavy army breakfast: eggs, bacon, SOS (Shit on a Shingle or, biscuits and gravy) and potatoes. Not a good load to be carrying on a run. The situation was further exacerbated by the fact that the food was seasoned with, to the American palate, strange Turkish seasonings. Not a good harbinger for a dazzling performance for the Brits.

After breakfast, the battalion formed up in a column of companies and Colonel Leighton gave the command, "Forward, march," and then, "Double-time, march."

The temperatures must have already been close to 90 degrees, and it wasn't long before soldiers started falling out of formation, doubled over, and throwing up. No one was spared. Officers and NCOs were just as susceptible as the junior soldiers. The once proud battalion formation degenerated into a straggling mass resembling Coxey's Army. All in all, it was a great embarrassment.

The British officers from the Parachute Regiment invited our officers to an entertainment to be held that night in the base officer's club.

"Did you hear that the Brits invited us to join them at the club tonight for some entertainment?"

"Yeah, that sounds cool. I wonder what it will be. I hope the club is air conditioned."

"I'll bet its belly dancing. We are in Turkey after all."

"You're probably right. I hope it is. I like belly dancing."

Full of high expectations, we arrived at the base officer's club at the appointed time. When all the officers had assembled, standing around loosely in the bar area, the show began. Rather than belly dancing, it appeared that the Brits had engaged a troupe from the nether regions of the Soho district of London. The troupe consisted of an MC, a short, roundish man with a thick Cockney accent, and three women. Two of the women looked to weigh 250 pounds apiece, but the third was very attractive; gorgeous, in fact, especially in comparison to the two behemoths.

One of the heavyweights began the show. Her routine involved performing acts with a teddy bear while buck naked. I'll spare the reader the details but suffice it to say that it elicited guffaws and a shower of pennies and ice cubes from the audience, particularly the Brits. Offended and angered by the lack of appreciation, she stormed out of the room, shouting, "I'm an entertainer," drawing even more hoots.

In the wake of her departure, there was a lull in enthusiasm on the part of the audience. Noting this, the MC pulled out his penis and started walking around the circle of spectators, wagging his member and exhorting us with raucous cries in his Cockney patter.

"'ello, 'ello chinas! Much more to come! Plenty more for your mince pies (eyes)! 'ello, ello!"

He repeated this performance a few more times, whenever he thought that enthusiasm was lagging.

The reaction to all of this highlighted differences between Brits and Americans, in spite of the fact that we share the same language (after a fashion). The Americans tended to be stunned, embarrassed, and somewhat revolted by the performance. The Brits, on the other hand, seemed totally unphased, treating each shocking spectacle as an opportunity for engagement. Evidence of this was the showers of pennies and ice on the first performer and attempts to light the MC's penis on fire with their cigarette lighters.

The performance by the second heavyweight is lost to memory, but that of the third, the beauty, is vivid.

She entered the arena stark naked, as had her predecessors, and immediately went into her routine, which was interactive. In a flash, she snatched Lieutenant Mc Carter's thunder whistle and put it between her legs at her crotch. The whistle was attached to a lanyard and the beauty was steadily pulling Lieutenant Mc Carter's head toward her crotch. The Brits were hooting and roaring their approval. For me, alarm bells were going off. Jeff Guild and I had been standing at the front of the crowd but, at the sight of this, I quickly faded to the back. *'Whoa, mama'*, I thought, *'no way am I going to end up out there.'*

After this grand finale, the great show came to an end. The jump was scheduled for the following day, 26 September, and we needed to get some sleep.

26 September dawned sunny and hot but, to our dismay, the jump was postponed for 24 hours due to high winds on the DZ. Another 24 hours to kill and try to beat the heat. Another weird American/Turkish breakfast to eat.

Reports from the DZ on the following morning were good and the airborne operation proceeded. Staff Sergeant Parish and I were the jumpmasters on our aircraft. Our duties as jumpmasters were to inspect each jumper on our aircraft prior to boarding, making sure that parachutes and equipment were correctly worn and that there were no obvious safety issues.

During flight, once the pilots alerted us that we were 20 minutes from the DZ, each jumpmaster would hang out of the open door of the plane, observing for hazards from other aircraft and watching for the approach of the DZ. When the pilots gave the six-minute warning and switched on the red light by the door, the jump commands began. These had to be shouted in order to project over the roar of the engines and were accompanied by hand and arm signals.

"SIX MINUTES"

"GET READY"

"OUTBOARD PERSONNEL, STAND UP"

"INBOARD PERSONNEL, STAND UP"

"HOOK UP"

"CHECK EQUIPMENT"

At this command, each jumper would check the parachute pack of the soldier in front of him to make sure he was correctly hooked up and that there were no issues with the static line.

"SOUND OFF FOR EQUIPMENT CHECK"

When the equipment check was complete and no problems noted, each jumper, from the rear to the front, would smack the butt of the jumper in front of him and yell "OK". The last jumper, who would be the first to exit the aircraft, would point to the jumpmaster and yell "ALL OK". Then came the command,

"STAND IN THE DOOR."

When the plane was over the DZ, the pilot would flip on the

green light and the jumpmasters on each side of the aircraft would slap their respective jumper on the butt and yell "GO".

This sequence unfolded in textbook fashion. We were just about at the last jumper and preparing to jump ourselves when the green light went out and the red light came on, meaning the plane had cleared the DZ and no one else was to jump. This was only my fourth jump as a jumpmaster since completing the jumpmaster school, and I was going by the book. I stopped the last jumper and signaled SSG Parrish to do the same. Had I been more experienced, I probably would have ignored the red light and jumped anyway, but there we were, stuck on the plane and begging the pilot to make another pass, which he refused to do. We had to land with the plane and be ignominiously trucked up to the exercise area.

I have very little recollection of the exercise itself except for events at the end. All I remember is driving through villages with the men sitting there drinking tea or coffee, no women in sight. From what I have read since, the women were out in the fields working.

Colonel Meloy had directed that every soldier would have a steak dinner at the end of the exercise. His S-4 Supply Officer moved heaven and earth to procure that many steaks, get them shipped to Turkey, cooked and distributed to the scattered units in the field. Shortly before the steaks were to be served to C Company, my commander, Captain Pete Tolley, had to go to some ceremony on the spot where the airborne forces had linked up with the Turkish ground forces.

My pack, containing my poncho, extra socks, rations and toilet articles, was on the back seat of his jeep. The day was sunny and mild, not a cloud in the sky, so I thought nothing of leaving the pack in his jeep. After all, he wouldn't be gone long, and I had no need of anything—or so I thought. Just as the steak dinner was ready to be served, a storm cloud appeared over us, and it started to rain. Hard, cold rain. All the soldiers immediately wrapped themselves in their ponchos. There was no cover anywhere, and I had no choice but to stand there and get soaked to the skin. Cold and miserable, my soggy steak dinner was a letdown. I learned once again the old army adage, "Never get separated from your equipment."

At the conclusion of the exercise, the battalion was to be transported by bus to a bivouac area at Cakmakli. Major Riddick, the battalion S-3 operations officer, had meticulously organized the buses into a convoy. There was an order of march, time schedules, checkpoints, chalk numbers, and bus commanders equipped with radios. The only problem with this immaculate plan was that the drivers were Turkish and had not been read in on the plan. In their minds, they knew their destination and how and when they got there was their business. Order broke down immediately as buses overtook each other. Some drivers stopped to pray. Others stopped to purchase a snack or to say hello to an acquaintance. The finicky Major Riddick was pulling his hair out, but those of us who were bus commanders thought it was a hoot. In between messages to him intended to sooth his jangled nerves, we were announcing the progress of our convoy as if it were a horse or car race.

"Number 4 is in the lead. Number 2 has dropped out for prayer. Hello, what's this? Number 7 is coming up fast on the inside. Number 7 is closing in on Number 4, Number 7 is in the lead, ladies and gentlemen, Number 7 has taken the lead." And so on. Major Riddick was not amused.

We all arrived at Cakmakli safe and sound, if not on Major Riddick's schedule. Our bivouac area was spacious, with GP Medium and GP Large tents set up to house the soldiers and GP Small's for the officers. Passes to Istanbul were organized but, as per Colonel Meloy's orders, no soldier could go on pass without a condom. As OIC Condoms, I set up my Pro Station in the company area, sitting behind a field table with my box containing 144 prophylactics. It wasn't very exciting duty, but it did give me a chance to banter with the soldiers as they reported to me for their pass and protection. My opportunity for a pass would come soon enough.

There were two well-known houses of prostitution in Istanbul. One was called The Compound and was, allegedly, a prison where the inmates did penal servitude as prostitutes. The other, whose name escapes me, catered to a much higher class of clientele.

Many, maybe even most, of the soldiers would choose to remain chaste, but there was no doubt that others would patronize one or the other of these establishments. Rumor had it that Colonel Meloy had visited both, not as a customer but to assure himself that it would be reasonably safe for his soldiers to go there.

I saw one of our young soldiers after his return from his pass. "How was you pass, Butner? Where did you go?"

"Oh, hi sir. I went to The Compound."

"Did you use your rubber?"

"No, sir."

"What!? Why not."

"She told me she didn't want me to."

"For cryin' out loud, Butner, she didn't want you to? I hope you don't get a dose of the clap or something worse."

You can lead a horse to water, but you can't make it drink.

My turn for a pass came up. I had a first cousin, considerably older than me, who was married to a Turk who was president of the Bosporus University. I contacted her and we arranged to meet in Istanbul. I invited Lieutenants Steve Roy and Dave Hemingway to accompany me since they were interested in seeing the cultural sights of Istanbul. We met at the Dolmabahce Landing on the Straits of the Bosporus.

Catching a cab, which were almost uniformly big, fat, 1950's era automobiles like De Soto's, Hudson's, and Pontiacs, we saw the sights. We visited the Blue Mosque, the Hagia Sophia Grand Mosque and many other famous sites. The highlight for us, however, was the Grand Bazaar, where my cousin Sylvia turned us loose. Miles of sub-terranean tunnels featuring every kind of shop one could imagine. It was here that we experienced a brand of hospitality unknown in the United States. Our close haircuts and general physical appearance clearly identified us as soldiers. As we passed a shop selling oriental rugs, we were hailed by the proprietor. "Come in, come in. You must have a coffee. Would you like a coffee?"

He seated us in his shop and served us a Turkish Coffee while he questioned us about how we liked Istanbul and our time in Turkey. He knew we were soldiers and that we weren't going to buy a rug. Rather, he was merely showing hospitality. We had been apprehensive about wandering the streets of this exotic city, imagining that danger lurked on every street corner and alley. After experiencing it, we all agreed that we felt safer on the streets of Istanbul than on the streets of big cities in the United States.

We did do some shopping for souvenirs. I bought an embroidered sheepskin vest for my son, earrings and copper pans for my wife.

I had another opportunity to visit my cousin at her house near the university. I well knew her mother, my aunt, but had never met her before. It was a wonderful opportunity to meet more of my extended family. She is in her 99th year and we remain close to this day.

There were many other sights, sounds, and smells to remember.

Like riding in convoy in 2 ½ ton trucks, passing Gypsies camped along the side of the road in tents made of clear plastic and completely covered with flies; thousands upon thousands of them. As our trucks passed, clouds of flies would detach themselves from the Gypsies and hover at the rear of the trucks. None of us had ever seen anything like it.

Then there were the suicidal driving habits of Turkish drivers, but that perhaps should be left for another story.

Exercise Deep Furrow came to an end and we re-deployed to Fort Bragg, far richer for the experience and happy to have seen the world the poor man's way, the soldier's way.

Author in Incirlik, Turkey with American and British paratroopers

11. This Is The Real Deal

"Mission: At 0100 hours, 25 October 1973, 2-508 conducts a withdrawal not under enemy pressure and relocates to positions further to the rear. Commencing at 0100, platoons will start thinning the lines and reassemble at this location. The mortar section and one squad per platoon will be left behind as the Detachment Left In Contact (DLIC). Lieutenant Sherwood, you will be in charge of the DLIC. Once we are in the new position, the DLIC will be withdrawn."

My unit, the 2d Battalion, 508th Parachute Infantry Regiment, 82d Airborne Division was on its intensified field training cycle. In the 82d Airborne at that time, there were three cycles which rotated through the three brigades of the division. At any one time, one brigade would be training, one brigade would be on some level of alert and one brigade would be performing support functions around Fort Bragg.

At this phase of our training, the battalion was in the defense, facing a simulated enemy but not actively fighting. Withdrawing in the face of the enemy may be ordered for a variety of reasons but the important feature of such a move is that it be concealed from the enemy. Hence, withdrawals are ideally conducted under the cover of darkness and a detachment is left behind to create the impression that the entire unit is still in position. At the company

95

level, this task is normally given to the Executive Officer (XO). At this time, I was the XO of Company C.

Given this assignment, I made plans to spread the troops under my control across the entire company front and to simulate radio transmissions that would be typical if the entire company was present. The mortars were to fire intermittently. As the bulk of the company withdrew, my DLIC expanded to cover the company front. We continued our charade while waiting for the CO, Captain Pete Tolley, to give us the signal to withdraw.

THURSDAY, 25 OCTOBER 1973, 0300 HOURS:

"Charlie Five this is Charlie six, over."
"This is Charlie Five, over."
"This is Charlie Six, execute, over."
"Charlie Five, WILCO."

This was my signal to withdraw the DLIC and rejoin the company in their new location. I had the mortars fire a final fire mission and then began quietly withdrawing everyone back to the assembly point, which was located where the company command post (CP) had been. I had at my disposal my jeep and four mechanical mules[5], which would help speed our movement back to the new company position.

We would be travelling cross-country through the woods at about 3 mph with lights out. It would be slow going, but faster than if we all walked, lugging the mortars and other heavy equip-

---

5. The M274 Mechanical Mule was a motorized platform which was approximately 9 feet long, 3 ½ feet wide, 3 ½ feet high.

ment. I piled as many soldiers as would fit on my jeep and the four mules. Those that wouldn't fit walked along, with one in front of each vehicle as a ground guide.

We had travelled for about thirty minutes when I was shocked by the soft sound of bodies thudding against the ground, the metal on metal sound of mortar tubes, baseplates and bi-pods banging against each other ... one of the mules had tipped over.

My heart was immediately in my throat, with visions of crushed and maimed bodies.

"Shit! Turn on the lights!"

Tactical security gave way to what I feared was a medical emergency. As the lights of the jeep came on, I saw what I expected to see. Equipment and bodies were strewn to the side of the mule, which was turned over on its side. Running frantically to the scene, it began to dawn on me that it wasn't as bad as I feared. In fact, no one was hurt and none of the equipment was damaged. After accounting for everyone and all equipment and righting the mule, we resumed our tactical, black-out movement.

THURSDAY, 25 OCTOBER 0400 HOURS:

We rolled into the new company position at 0400. I reported to Captain Tolley and was in the process of returning the DLIC elements back to their parent platoons when the company radio crackled.

"Charlie-six, this is Sierra-one four, over."

It was Smitty, our Support Platoon Leader.

"This is Charlie-six, over."

"The exercise is terminated. Assemble your unit for movement to the rear. We are on alert. This is the real deal. I say again, this is the real deal. My element will immediately begin shuttling subordinate units to the rear, over."

Captain Tolley's response was as simple and brief as it could be. "WILCO."

When we first arrived at the new location, I was dead tired and looking forward to getting a little sleep. Now I was electrified. I had no idea what we were being alerted for but knew it had to be serious. I was filled simultaneously with a sense of dread and exhilaration. Those who have served in the combat arms would probably be hesitant to admit it to those who haven't, but there is a certain lust for action that runs counter to the rational desire that there be no conflict. The closest analogy I can make is that of a contact sport athletic team that constantly trains for a game but never gets to play. When the opportunity to play comes along, there is a real excitement about getting out on the field and trying it on with one's opponent.

Captain Tolley broke me out of my revery.

"Buzz, I want you to organize the movement. Figure out the order of march, the ACL for each truck, etc. First Sergeant Roach will work with the platoon sergeants to get the troops organized for the move."

"Yes, sir."

The battalion had approximately six hundred troops in the field and Smitty had eight 2 ½ ton trucks with which to shuttle everyone back to the rear. My task was to figure out how many trucks would be needed to move our approximately 125 troops and to organize the truck loads. Figuring we could cram 24 or 25 standing soldiers onto each truck, I calculated we would need five trucks to move the whole company. The First Sergeant began organizing the truck loads while I contacted Lieutenant Smith.

"Sierra one-four, this is Charlie five, over."

"This is Sierra one-four, go."

"Roger. We need five trucks to move this element. We'll pack the bulky equipment and weapons in my trailer and on the mules, over."

"Roger. I'm moving Alpha and part of Bravo right now. As soon as I drop them off, I'll send five trucks to your location. ETA approximately one hour, over."

"Roger. We'll be standing by and ready to move."

Meanwhile, Captain Tolley left to go to the rear to meet with Lieutenant Colonel Pat Leighton, our battalion commander, to get briefed on the situation and receive orders.

Having gotten the company organized into truck loads, there was nothing to do but wait and speculate.

"Whaddaya think's up?"

"Hell if I know, but it must be serious."

"It probably has to do with the war in the Middle East."

"That sounds about right."

"Ooh, that would be some serious shit."

"You got that right."

"Well, we'll find out soon enough."

The young soldiers were visibly excited. Those who had been to Vietnam less so. In fact, the veterans were decidedly sober.

0600 HOURS, 25 OCTOBER, 1973

Smitty's trucks arrived in due course and transported the company to the rear. When Captain Tolley returned from meeting with Colonel Leighton, he gathered his leaders together.

"Here's the skinny. The whole division has been alerted to intervene in the Mid-East war. Right now it looks like an airborne operation. The IRC and the DRF 1 are already down at Green Ramp drawing parachutes and ammunition. Since we've been on training cycle, we will be in the follow-on echelon. Right now, we need to get A-bags packed and vehicles rigged for heavy drop. Weapons won't be issued until we get a movement time. If there's time, married personnel will be allowed a quick trip home to say goodbye and secure any needed uniforms or equipment. If there's no questions, let's get moving."

What happened next was truly awe inspiring. This unit, which had been training in the field for several days with minimal sleep, now completely changed course and prepared to be deployed into a combat zone, to face who knew what. Soldiers who had, short hours ago, been practicing for combat, now packed their bags to be parachuted into combat. Vehicles which had, short hours before, been driving through the woods, now had to be rigged to be dropped out of a plane by parachute...and all in a matter of a few short hours. Everyone went about their business in a totally professional manner, as if this was what happened every day. Soldiers who were on leave voluntarily came back. Even soldiers who were AWOL returned, not wanting to miss the real deal. I was even able to make a quick trip home to pick up some equipment and to say goodbye to my wife. It was all so sudden that there wasn't even time for anguish and fear. That could come later.

1600 HOURS, 25 OCTOBER 1973

The battalion was ready to go well within the prescribed time. Now we stood by and awaited our turn to move to Green Ramp to draw parachutes and ammunition and board planes bound for, word had it, a jump onto the Golan Heights. With adrenaline now subsided we now had time to ponder soberly what awaited us. Any euphoric excitement was replaced by a grim determination to do our duty and live up to our reputation.

That night, officers and NCO's who lived away from the barracks bedded down on the floor in the various offices, anywhere there was space to stretch out. All of us slept with our gear at our side, ready to move out on a moment's notice.

26 OCTOBER 1973

With dawn came the word that the alert was called off. It appeared that our readiness to insert ourselves between the Syrians and Israel was sufficient to make the Syrians think twice about their plan to invade.

That same day, our division commander, Major General Frederick J. "Fritz" Kroesen, wrote a congratulatory letter to the members of the division. It is worth reproducing it here in its entirety.

This was a little over a year since I had returned from Vietnam—a little over a year since the disengagement of U.S. combat forces from Vietnam after seven years of fighting there. It was said that the Army was in a shambles, that morale, discipline, combat readiness, and every other indicator of military efficiency was at the lowest possible ebb. That may have been true in other parts of the Army, but it wasn't true in the 82d Airborne Division. I could not have been prouder.

DEPARTMENT OF THE ARMY
HEADQUARTERS 82D AIRBORNE DIVISION
FORT BRAGG, NORTH CAROLINA 28307

AFVCCG   26 October 1973

SUBJECT: Deployment Alert

TO: All Troopers, 82d Airborne Division

1. On 25 October the 82d Airborne Division was alerted for possible
imminent deployment to the Middle East. At the time our mission and the
role we were to play was unclear, but I assure you the alert was real, the
reasons were valid, and we were being called upon to demonstrate that we
are a truly ready, professional combat force.

2. Initially preparations were made for an airborne combat assault, and
elements of DRF #1 began rigging equipment for heavy drop, parachute packing and issue plans were stepped up, and tactical planning was begun.
Later, when additional instructions were received, plans were changed and
we prepared for an airland operation. Much of the hard work accomplished
for an airborne assault had to be undone, and more hard work had to recon-
figure the equipment loads. I apologize to the men who had to do this
double work, but I want you to know it was because of conscious decisions
on my part, it was not because of confusion, and given the same kind of
alert next time we will do it the same way.

3. There is no question about the readiness of the 82d. Your demonstration on Thursday of your ability to load up and move out was watched closely
and reported by members of the press and television cameramen. Their stories
reflected the respect you generated among these men, and you know that they
are men who are not easily impressed. One said that you went about your
business "calmly and coolly with a deadly precision that would make a
division commander smile." You may not have seen me smiling, because our
business can truly become deadly and not a smiling matter, but I assure you
that I was proud to be known as a member of this division on that day. I
hope you were too.

4. Most importantly, your readiness may have been one of the significant
contributions to the cooling off of a very threatening situation. Without
the 82d standing by, obviously ready to take off, it is questionable whether
or not our government's resolution to keep the super-powers out of the Middle
East would have been adopted by the Security Council of the United Nations.
We can't take credit for preventing a war, but we know our readiness might
have caused some second thoughts among those who were weighing the risks.

REPORTING FOR DUTY

AFVCCG

26 October 1973

All Troopers, 82d Airborne Division

5. Many of you were called back from leave, from TDY assignments, from schools. Your response was great and I want to assure you that we'll get you back to school, back on leave if you wish, and that you will be reimbursed for any travel expenses incurred because of your recall. (Hang on to your leave authorization, DA Form 31, so you can claim your money.) *make sure they return leave form*

6. I have made many statements in the past that our mission is combat readiness. You proved on 25 October, in fewer hours than many thought possible, that we are accomplishing our mission outstandingly. I hope this occurrence gave you a clearer understanding of why this is our mission and why it is significant to our nation and the world. This time we didn't have to go, and maybe your life or my life was saved because of that. For next time, we have to be just as prepared.

FREDERICK J. KROESEN
Major General, USA
Commanding

12. Welcome to the Big Red One

1975 was an eventful year. I was a newly promoted captain attending the year-long Infantry Officer Advance Course at Fort Benning, Georgia. In April of that year, Saigon fell to the North Vietnamese Army, effectively stranding from their homes and families the many South Vietnamese officers who were attending the course. Aside from our feelings of sympathy for our fellow students who no longer had a country to return to, the fall of South Vietnam didn't cause much hand wringing or sadness. For us Americans, the war had effectively ended in 1972 and was no longer our problem or concern. This seems odd, given the enormous sacrifices made by U.S. soldiers in that conflict. The Army hierarchy seemed content to put Vietnam in the rear-view mirror, focusing instead on a more conventional war against the Soviet Union.

It was also an eventful year in ways much more personal. In May we would be receiving orders for our next assignments, dictating where we would be spending the next three years of our lives. Would it be Fort Polk, located in the backwoods of western Louisiana, or perhaps Fort Lewis, located south of Seattle in the scenic Pacific Northwest? The Army had just formed three new divisions, one at Fort Stewart, Georgia, one at Fort Polk, Louisiana, and one at Fort Ord, California. These seemed likely recipients of this new crop of captains coming out of the Advance Course.

I very much wanted to go to Fort Lewis, Washington and made my wishes known to the Infantry assignment officer. One place I was convinced that I wouldn't be assigned was Fort Riley, Kansas. Whenever my wife expressed dissatisfaction with our assignment, I reminded her that it could always be worse, we could be stationed at Fort Riley. I repeated this mantra so many times that Fort Riley became a bogey man, not only for my wife, but for me as well. I'm sure, dear reader, that you can see where this is heading. This is the way it happened:

Student companies at that time were two hundred strong, and many of our classes were held in large, two-hundred-man classrooms. We were seated according to our roster number. On the fateful day when we were to receive our orders, we filed into the classroom and there, at our assigned seat, was our orders—or not. 1975 was also a year in which the Army was scaling back the officer corps from its bloated war-time strength in what was referred to as the Reduction in Force (RIF).

Accordingly, at each seat was either a set of orders for the next assignment or a RIF notice telling the recipient that his services were no longer needed or desired. Everyone found out all at once, with no privacy or opportunity to compose oneself. To say it was an awkward situation is a vast understatement.

When I beheld my orders for the 1st Infantry Division at Fort Riley, I almost fainted. My own, self-created nightmare had become a reality. After regaining some semblance of composure, I began asking my friends about their assignments, and a pattern began to unfold. It appeared that the Infantry assignments officer, rather than trying to match assignments to expressed preferences, merely assigned people alphabetically. Those with last names beginning with A-M had orders for Fort Lewis. The rest of us were scattered among the other posts needing officers. That Dan Schilling and I received orders for Fort Riley suggested that the S's were next in line when the requisitions for Fort Riley were being filled.

Now that my own worst nightmare had come true, I had to tell my wife that hers had as well. She, being in New Orleans staying with family, would have the news conveyed to her over the phone. This was even worse than telling her that I received orders for Vietnam, since in that case I had told her before we were even married that I would, in all probability, be going to Vietnam. Back home and waiting to make the call, I broke out in hives for the first and only time in my life. The pain and itching were so bad that I finally took myself to the dispensary, where I was given a shot to alleviate the discomfort.

In spite of much drama and threats of divorce, in August the car was packed and our family of four headed for Fort Riley. The directions to get there from New Orleans were quite simple: head north on I-55 and hang a left onto I-70 at St. Louis, Missouri. I had only two impressions of Kansas and neither was good. One was the images of Kansas as portrayed in the Wizard of Oz and the other was the General Mills cereal commercials shown on the Lone Ranger TV program, showing flat fields of waving wheat as far as the eye could see. As we drove west on I-70, I kept a sharp eye out, noting the diminishing number of trees and the gradual flattening of the terrain. I was praying that we would arrive at Fort Riley before Kansas matched the images in my mind.

My prayers were answered. Fort Riley sits in an area of Kansas known as the flint hills. We found a nice bungalow house in the old part of Manhattan, home to Kansas State University. Manhattan proved to be a lovely place to live—small town America at its best.

On reporting to the 1st Infantry Division, I discovered that I was earmarked for the 1st Battalion, 28th Infantry, known as the Black Lions of Cantigny. It was commanded by LTC George Geczy, a friend of my former commander in the 82nd Airborne Division, LTC Pat Leighton, who had arranged for me to be assigned there. This turned out to be fortuitous in many ways, initially because the battalion had the reputation of being a good unit.

After being welcomed to the battalion, I was assigned to the S-3, the training and operations section under Major Al Wissinger, an infantryman of vast experience from whom I learned much. This was exactly where I wanted to be—except for the fact that there were four other captains and a lieutenant in a section that called for three officers instead of seven. It was like a parking lot with more cars than parking spaces. One had to arrive extra early just to claim a desk, and there simply wasn't enough meaningful work for that many people.

Shortly after being assigned to the battalion, the other new officers and I were called to the brigade headquarters for an in-brief by the brigade commander. My Advance Course classmate, Dan Schilling and I were there from the Black Lions and there were a couple more captains from our sister battalions. All of us were recent Advance Course graduates, in top physical condition, close military haircuts, with crisply starched uniforms and highly polished boots.

We reported to the colonel and were told to take a seat. He was bald, tall and lean, almost skeletal. He had a high-pitched voice and a heavy southern accent. Rather than welcoming us or telling us how lucky we were to be members of the famous Big Red One, he launched into a diatribe about his pet peeves.

"I don't like fat officers or officers with long hair. And I can't stand officers who can't balance their checkbooks ..." and so on in a similar vein.

'Why is he talking to us like this,' I wondered. Each of us looked like we'd stepped out of a recruiting poster and there was no reason to suspect our professionalism, on duty or off. It was a very unpleasant introduction to the most negative command environment I encountered in my military career. I was soon to discover that the division commander and assistant division commander were equally negative. Lieutenant Colonel Geczy was a good commander from what I could observe, but he had a tendency to pass on to his subordinates whatever negativity came from above, giving credence to the old Army saying, "Shit rolls downhill."

All this was to change in early December when LTC Geczy relinquished command of the battalion to LTC Neal T. "Tom" Jaco. Colonel Jaco's arrival was like a breath of fresh air. We still had the same toxic division and brigade commander, but it seemed that part of Colonel Jaco's philosophy was to insulate and protect his subordinates from having to function under a cloud of fear. The cocoon he created wasn't impenetrable, but professional life improved markedly under his command.

Changes were taking place for me as well. In January, Colonel Jaco selected me to be the battalion Adjutant. I received this news with mixed feelings. On the plus side, I was flattered that the new commander was willing to entrust me with a primary staff position. On the negative side, since I considered myself primarily a field soldier, I fretted at being put in charge of the personnel administration of the battalion. Most of all, I feared that this assignment would prevent me from taking command of a company because I assumed I would have to be the Adjutant for at least a year. I needn't have worried. I was selected to take command of

Company A in March, a mere three months after becoming the Adjutant.

In spite of the negative command environment at division and brigade level, I was proud to be a member of the famous Big Red One. I was prouder still of our battalion. The Black Lions had always enjoyed a tremendous esprit de corps, but it soared to even greater heights under the leadership of Colonel Jaco. This was to be my home for the next three years, but in ways that I couldn't have possibly foreseen.

13. Frontier Justice

Fort Riley is located in north central Kansas in an area called the Flint Hills. Founded in 1853 on the western edge of what was then the frontier, it remained during the 19th Century a central headquarters for the Indian fighting army of the west. Vestiges of the frontier often remain in areas of the country that have become settled, but not fully urbanized. Such was the case when it came to justice for Private Joe S. De Ragman.

In March 1976, I took command of Co A, 1st Battalion, 28th Infantry, Black Lions, 1st Infantry Division, The Big Red One. At this time, the Army was struggling to re-instill discipline and esprit de corps after seven ravaging years of war in Vietnam. The famous Big Red One was no exception.

My first task on assuming command of the company was to clean out the drug users and hard-core thugs who terrorized the barracks at night. Having accomplished this, the next task was to ratchet up the level of discipline, thereby giving the soldiers a unit to which they could be proud to belong.

Joe De Ragman was not a drug user or a thug, but he was not a good soldier. His seeming inability to adhere to high standards was brought to my attention after his platoon chain of command had exhausted the resources available to them. He was habitually late to or missing formations and no amount of effort by his sergeants or lieutenant served to correct this situation. He was referred to me for company punishment.

Since missing formation is a relatively minor offense, I elected to begin at the lower end of the punishment options available to me. I chose not to administer an Article 15, which was the official form of punishment at the company level, becoming a permanent part of a soldier's personnel file.

"De Ragman, I'm going to go easy on you this time. Instead of an Article 15, which could involve reduction in rank and forfeiture of pay, I'm going to send you to School of the Soldier this weekend. Hopefully that will help you get your head screwed on right and make you realize that you are in the Army and are going to do what is required of you. In this man's Army, you don't get to decide what you are or aren't going to do. First Sergeant, make the necessary arrangements."

School of the Soldier was a mechanism put in place by higher headquarters to deal with just such minor infractions. Instead of enjoying a weekend of free time along with their fellows, soldiers sent to School of the Soldier spent the weekend doing physical training, drill, and other basic soldiering skills. It was meant to instill pride and discipline while at the same time being unpleasant enough that one would not want to do it again.

On the Monday following De Ragman's assignment to School of the Soldier, it was reported to me that he had failed to report there as ordered. This was a violation of Articles 86 and 92 of the Uniform Code of Military Justice (UCMJ): Failure to Repair and Failure to Obey a Lawful Order. Now I had no option but to deal with him through Article 15 of the UCMJ. I had the necessary paperwork typed up and, at the time prescribed, De Ragman reported to me for the proceeding. Present were his team leader, squad leader, platoon sergeant, platoon leader, and the First Sergeant.

After he reported to me, I began the proceedings.

"Private First-Class De Ragman, I am considering whether you should be punished under Article 15 of the Uniform Code of Military Justice for the following offenses:

"In that you did, on or about 5 May 1976, disobey a lawful order to attend School of the Soldier. This is a violation of Article 92 of the UCMJ.

"In that you did, on or about 5 May 1976, fail to report to School of the Soldier. This is a violation of Article 86 of the UCMJ. Before we proceed any further, I want to make sure you understand your rights under the Uniform Code of Military Justice.

"You are not required to make any statements, but if you do, they may be used against you in this proceeding or at a trial by court-martial. You have several rights under this Article 15 proceeding. First, I want you to understand I have not yet made a decision whether or not you will be punished. I will not impose any punishment unless I am convinced beyond a reasonable doubt that you committed the offenses.

"You may ordinarily have an open hearing before me. You may request a person to speak on your behalf. You may present witnesses or other evidence to show why you shouldn't be punished at all (matters in defense) or why punishment should be very light (matters in extenuation and mitigation).

"I will consider everything you present before deciding whether I will impose punishment or the type and amount of punishment I will impose. If you do not want me to dispose of this report of misconduct under Article 15, you have the right to demand trial by court-martial instead.

"In deciding what you want to do, you have the right to consult with legal counsel located at the Staff Judge Advocate's office on main post. You now have 48 hours to decide what you want to do."

"Do you understand everything I have just read to you?"

"Yes, sir."

"Do you have any questions?"

"No, sir."

"OK. Now we'll move on to the next step in the process. Do you want to seek legal counsel?"

"No, sir."

"Do you want to demand a trial by court-martial?"

"No, sir."

"Do you want an open or closed hearing?"

"What's an open hearing?"

"An open hearing would be open to the whole company. A closed hearing would involve only the members of your chain of command that are here now."

"I want an open hearing."

"OK. In that case, we will now stop this proceeding and resume it when we have the whole company assembled. We'll continue as soon as I can set it up. You're dismissed."

De Ragman saluted, executed an about face and marched out of my office.

This was new. I'd given several Article 15's and had never had a soldier request an open hearing. I was a little concerned. How would the company react? Discipline is about getting soldiers to conform to the norms of the unit. This appeared to be an open invitation for them to reject those norms. However, in spite of my trepidation, an open hearing was De Ragman's right, and he would have it, come what may.

A couple days later, the whole company was training down on the banks of the Republican River. That seemed as good a place as any to reconvene De Ragman's Article 15 hearing. After lunch I set myself up on a driftwood log and called the company together and explained the situation to them.

It was a clear, sunny day and the setting was rustic. I was sitting on the tree trunk with the river to my right. To my front was the gathered company, approximately one hundred soldiers, sitting or standing on the sandy bank. Off to the left were cottonwood trees, where the river bank gave way to higher ground. Right in front of me was the accused, PFC Joe S. De Ragman.

"Listen up! Private De Ragman was sent to School of the Soldier last weekend and he failed to show up. I've offered him an Article 15 for failure to repair and failure to obey a lawful order. One of your rights as a soldier if you are given an Article 15 is to have an open hearing. Private De Ragman was offered an Article 15 on Monday, and he requested an open hearing in front of his peers and that's what we are going to do now.

"So here's what has happened so far: I've read to him his charges and explained to him his rights — which are basically the military version of our rights under the 5th Amendment to the Constitution. He decided to accept the Article 15 rather than demand a trial by court-martial. As I mentioned earlier, he decided he wanted to have an open hearing and that's where we are right now. What's going to happen now is I'm going to turn it over to Private De Ragman and let him present his case."

Turning to him I said, "OK, De Ragman, the floor is yours."

I had no idea what De Ragman planned to say. Would he give an impassioned speech? Would he instigate a rebellion?

It immediately became apparent that he had given less thought to what he was going to say than I had. In a rather whiney voice, he said, "I didn't think I should have to do it. It's too hard."

De Ragman paused, seemingly at a loss for anything else to say.

The void created by his pause was immediately filled by a spontaneous eruption from the soldiers of the company. His feeble plea was met with guffaws, hoots and laughter.

"You pussy!"

"I went to School of the Soldier. It was easy."

"Ahh, can you believe that?"

"What a shit-bird!"

Whatever leverage De Ragman thought he would obtain from an open hearing had obviously not happened. He was a pathetic sight, standing there alone between his unsympathetic comrades and me.

Once the uproar had subsided, I asked him if he had anything further to present in the way of defense or mitigating circumstances. Head hanging and looking forlorn, he answered "No."

I proceeded to mete out his punishment, which was a reduction in rank and forfeiture of seven day's pay. This was perhaps overly harsh, but at the time and in the circumstances of the moment, it seemed appropriate.

A few days later, De Ragman went AWOL, never to be seen again. The company was better off without him and, hopefully, he was better off without the company.

On reflection, the outcomes all seemed to be in the plus column:

Justice had been served, rough as it was.

The scales seemed to be tipping in the direction of discipline and esprit de corps—the soldiers of the company seemed ill disposed toward tolerating shirkers and malingerers.

And, no soldier ever again requested an open hearing.

14. New Chain of Command

Commencing in June, the Big Red One got a completely new chain of command and the atmosphere of the division did a 180-degree turnabout.

Maj Gen Calvert P. Benedict took command of the division in June. I had previous exposure to him when he was Assistant Division Commander of the 82nd Airborne Division and already had a favorable opinion of him. He proved to be "just what the doctor ordered." He seemed to subscribe to the philosophy that if you expect good things, you get good things.

Virtually overnight from the moment he took the division flag, morale soared. Company commanders were no longer glued to their chairs in the morning waiting for the inevitable demands from division headquarters for reports on this, that, and the other thing. Instead, we were out where we belonged, running PT with our units. Why? Because the division commander was out there leading the way.

Brigadier General John Johns became the Assistant Division Commander. Unlike General Benedict, he was not a highly decorated combat commander, but he brought to the division other skills. Instead of relying on screaming and terror, he provided a level-headed focus on leadership and organizational efficiency. Under his tutelage, every soldier in the division from captain to corporal attended the Leadership Management Development Course (LMDC), some of the lessons of which I vividly remember to this day.

In the years that I have been retired from the Army, I have observed or experienced many so-called innovative management, leadership, and planning programs, the authors of which were quite proud of their novel ideas. Most of these programs had been part and parcel of the Army for years, in some cases as long ago as WWII.

Our new brigade commander was Colonel Jack Nicholson. His arrival was preceded by much fanfare about him having authored the new and much awaited Field Manual (FM) 100-5, Operations. This FM contained all the latest doctrine on field operations, especially the conduct of the air-land battle which would be conducted against the Red Army, should it come to that. Col Nicholson was said to be the guru of this new doctrine. Whether he was or not I never knew, as I never operated under him in the field nor did I detect any major changes to our training.

In any case, he, like the new commanding general, brought a positive energy to the unit. Two things I did know about him: he liked to have fun, and he liked to run. One of his favorite pranks was to take his three battalion commanders out for an evening and get them good and drunk. Then, bright and early the next morning, he would collect them and take them on a horrendous run.

The first outing of this sort was an unpleasant surprise. The next time Colonel Nicholson had a drink fest, the battalion commanders knew what to expect the next morning. Colonel Jaco, while not looking forward to it, was at least game. Less game was the commander of one of our sister battalions, who was found hiding in a toilet stall in the latrine.

15. Barbecue

Fort Riley may be in the Flint Hills, but it is still out on the prairie. It's extremely cold in the winter, extremely hot and dry in the summer, and the wind blows all the time. What few trees that exist are found by the rivers. The rest of the ground is covered with prairie grass or hay. In the summer, the heat, dryness and wind made wildfires a constant concern. One spark from a smoke grenade or even a blank round could start a grass fire. Every armored personnel carrier (APC) carried several "fire beaters", which were long poles with a thick, rubber mat attached to the end. The idea was to immediately beat out a fire as soon as it started and before the wind could catch it. The other fire control tactic was to race downwind in the APC's and cut a fire break before the fire got that far.

Sometime in the late summer of 1976, Company C, under Captain Jim Hosel and Company A, under myself, were maneuvering against each other. A fire accidentally started, caused by either a smoke or CS grenade, and quickly got out of control. It burned a considerable number of square acres before we could put it out, using a combination of the above methods.

A couple of days later, I was notified that Maj. Gen. Benedict was going to fly out to inspect my training. I was nervous about what his reaction would be to the fire, but other than that, I was satisfied with the quality of our training and was prepared to brief him. At the prescribed time, I was waiting for him at the map co-ordinates I had provided, ready to pop a smoke grenade to signal where his helicopter was to land. As a precaution against sparks, I carefully cleared a patch of ground of any grass and, as a further precaution, I held the grenade aloft instead of throwing it on the ground.

As General Benedict's helicopter landed, I ran forward to greet him as he stepped off. "Black Lions, Sir! Captain Sherwood, A Company."

As he returned my salute, I led him to the site I had set up to brief him. I showed him on the map the disposition of my pla-toons and explained our training objectives.

"Sir, we've been conducting an opposing forces operation with Charlie Company. We've been practicing platoon in the attack and defense. The focus in the attack has been on fire and movement, and overwatch techniques. In the defense, we have focused on ter-rain selection, weapons placement, and defense in depth."

Seemingly satisfied, General benedict said, "Let's go up and have a look."

Boarding his helicopter, the crew chief handed me a headset so that the general and I could talk over the intercom system.

As we flew around the training area, I pointed out the locations of my platoons and explained what was going on at present, as well as what we had been doing over the preceding days.

A major distraction was the acres and acres of blackened land below us and, of course, I had to explain the fire and the measures we had taken to get it under control. Near the end of our flight, we came upon a huge haystack, comprised of hundreds of bales of hay, all scorched black by the fire. As we passed over, the general said to me over the intercom, "I hope the cows like barbecued hay."

16. Naughty Boys

The draft Army and the earliest years of the volunteer Army was nothing if not made up of characters. A less euphemistic description would be troublemakers.

There are different kinds of troublemakers, and each can be handled in different ways. One kind of troublemaker is the kind that seeks to disrupt the integrity of the unit by refusing to obey orders, constantly challenging authority, or even disrupting the sense of safety in the unit by terrorizing fellow soldiers during after duty hours. This type of troublemaker needs to be weeded out quickly for the good of the unit.

The other type of troublemaker is more like a grownup version of naughty schoolboys who can't stay out of trouble and are regular visitors to the school disciplinarian. They mean no harm to the unit or their fellow soldiers, they just find it impossible to behave themselves all the time.

For me, as a company commander, the former type was to be eliminated from the service as expeditiously as possible. The latter type was to be tolerated. No transgression could go unpunished but if, in my assessment, they would be there and fight for the unit when the chips were down, there was nothing to be gained by eliminating them. A religious commander might have described them as his cross to bear.

In A Company of the Black Lions, four such characters stand out. Like the naughty boys who always collect in the back of the classroom, they gravitated to each other and created a thorny foursome. All four were from New York City. They were named Walsh, Fitzpatrick, McMonigle, and Vianney.

Walsh, the shot-caller of the group, was a big hulk of a man with potential later in life as a bouncer or perhaps an enforcer in the Irish Mafia. If Walsh was the Victor McLaglan of the group, Fitzpatrick and McMonigle were the Barry FItzgeralds. Small in stature and full of nervous energy, they were ever alert for opportunities for mischief and to do Walsh's bidding. Vianney was short and roundish. One could picture him later in life with an ever-expanding waistline propelled by beer and fatty food. He was of an easy-going nature and was happy to go along with his Irish cronies.

A story might best illustrate the type of mischief that naughty boys get up to in an Army unit. Following is an incident that is, as they say in the movies, based on true events. The essential facts are true, enlivened by my imagination as to how it might have gone down.

Each battalion had its own motor pool containing all its tracked and wheeled vehicles. In an infantry battalion, there were over one hundred tracked vehicles plus numerous jeeps and 2 ½ ton trucks. In the mid-1970's, the M113 Armored Personnel Carrier (APC) was the primary vehicle in a mechanized infantry unit. It was basically an armored shoebox on tracks designed to carry a ten-man squad.

The rear of the APC had a ramp which lowered hydraulically. Within the ramp was a smaller crew door which was locked from the inside. On the top of the APC were two hatches, one in the left front corner for the driver and one in the center for the vehicle commander. When parked in the motor pool, these hatches were secured with a padlock.

Because of the sensitivity of all this expensive equipment, the motor pool was guarded during off-duty hours by soldiers of the battalion. Included in the guard's duties was the requirement to check each vehicle to make sure it was properly secured. Any irregularities were to be reported to the Sergeant of the Guard.

On the night in question, Vianney and Fitzpatrick were on guard. Together, they were checking the security of each APC, one shaking the padlocks on the hatches and the other testing the handle on the rear crew doors. About half-way through the process...

"Hey, Vianney, this door is unlocked."

Hopping down from his perch where he was checking the hatches, Vianney said, "Oh, wow! I guess we ought to tell Sergeant Dye."

"No, let's go in and see if there's any good shit laying around."

Climbing into the passenger compartment, they turned on their flashlights for a look. "There's nothing here. You got any cigarettes?"

"Yeah, here."

Shaking a couple out of the pack, they both lit up. "I've always wanted to drive one of these. Do you know how?"

"A little. Sergeant Reynolds tried me out as a driver, but I guess I pissed him off or something."

"C'mon, let's drive around the motor pool. It won't hurt anything. No one will find out. You can show me how."

Vianney crawled up into the driver's compartment and started the engine. Fitzpatrick stood in the TC's position. Since both the hatches were padlocked, the only field of vision for either was through the viewing slit in the hatch and it was the middle of the night. For those with no experience in an APC, imagine trying to drive your car at night with no headlights, no streetlights, and only a two by twelve inch slit at the very top of your windshield to see through.

After letting the engine warm up, Vianney put it in reverse and backed around into the lane.

"Lemme drive, lemme drive!"

"Wait a minute. I'm just going to drive around the motor pool a couple of times, then you can drive. Watch me."

In spite of the dark, the silhouettes of the other APC's were discernable, which allowed Vianney to guide on them. When he reached the end of the row, he pulled back on the left lateral, making a left turn along the perimeter of the motor-pool. After two more left turns along the fence line, he was now approaching the back row of vehicles where the jeeps were parked. The jeeps, being less than half the height of the APC, were invisible to the driver. Yelling over the engine noise, Vianney yelled, "Fitz, I can't see, can you?"

"I think we're at the end. Turn 'V', turn!"

Responding to Fitzpatrick's instruction, Vianney pulled back on the left lateral and, as the APC turned to the left, both felt rather than heard a crunching and a jarring of the APC.

"What was that?"

"I don't know. I think we hit something."

"Stop! Stop!"

Taking his foot off the gas and pulling back on both laterals, Vianney brought the APC to a halt.

"Fitz, go see what happened."

Fitzpatrick exited through the crew door and went around to the front of the APC.

"Fuck, it's a jeep!"

Jumping up on the sloped front of the APC, Fitzpatrick shouted through the driver's hatch. "V! We ran over a jeep!"

"Shit! What are we going to do?"

In a very adolescent strategy, these two characters decided to re-park the APC and go about their business as if nothing had happened, hoping that no one would notice, or at least not suspect them as the culprits. As with most adolescent evasion of guilt plans, their "I don't know what happened" alibi quickly fell apart under first scrutiny.

This was a very serious and expensive offense which could have easily merited a court martial or a less than honorable discharge from the Army. On the other hand, their mischief notwithstanding, these were trained soldiers who were loyal to the unit and could be expected to be there when it was time to do the business. What was to be gained by sending them to prison or throwing them out of the service?

In consultation with Colonel Jaco, he and I shared the same view. The offense was referred to a battalion level Article 15 because the offense merited more punishment than I as their company commander could impose. Colonel Jaco hit them hard, reducing them to the lowest rank and fining them half a month's pay for two months. Additionally, they were financially liable to pay for the damage to the jeep.

In a case of justice tempered by mercy, Fitzpatrick and Vianney went on to deploy with the battalion to Germany, where they continued to serve as loyal but mischievous members of Alpha Company of the Black Lions.

17. Paint It Black (Lion, That Is)

I wanna see it painted
Painted black
Black as night
Black as coal...

—Mick Jagger and Keith Richard

As commander of Company A, 3d Battalion, 28th Infantry, 1st Infantry Division, I assembled my subordinate leaders in my office and began to issue an operations order.

"OK. This is the situation, 3d Battalion, 28th Infantry departs Fort Riley at 2400 hours, 12 November enroute to the Federal Republic of Germany. Mission: Company A, 3-28 Infantry commences operations at 1900 hours, 12 November 1976 to paint Black Lions all over Fort Riley as a special farewell gesture."

Before continuing with this saga, it's necessary to go back several months and recount the events leading up to this moment. In the Spring of 1976, the 1st Battalion, 28th Infantry received notification that it would be moving permanently from Fort Riley, Kansas to Wiesbaden, Germany in November to replace the 1st Battalion 18th Infantry as part of Brigade 76.

For me and for most of the members of the battalion, this was good news for a number of reasons.

Service in Germany seemed very attractive because of the opportunities to travel throughout Europe.

As a professional soldier, I liked the idea of being on the front line of the Cold War, with a real mission to defend Western Europe against the Soviet Union.

Additionally, after suffering with antiquated, gasoline-powered armored personnel carriers (APC) which seldom ran, and for which we could not get repair parts, the prospect of drawing new, functioning, diesel-powered vehicles when we got to Germany was truly exciting.

Moving a battalion of 800 men is a major undertaking. This was especially so for us, as the move was a permanent one, not a temporary six-month deployment. There was much to be done, some of which seemed to work at cross purposes.

There was weapons qualification and extensive field training to be accomplished which would culminate in a tactical evaluation of the battalion to ensure we were ready to play our role as part of U.S. Army Europe and NATO. At the same time, we had to prepare our worn-out armored personnel carriers (APC) and wheeled vehicles for turn-in, a very arduous process which was not enhanced by rigorous field training.

Most of the summer would be taken up with support of ROTC summer camp, running ranges, and generally assisting in the training of officer cadets. For unit commanders, this task was seen as an unwelcome distraction from our own training and ability to prepare for our overseas movement. Except, that is, for the positive impact it had on the morale of the troops.

This was an opportunity for junior ranking soldiers to shine in the eyes of the cadets. They were real soldiers and often knew more about the profession of arms than did the cadets. Every opportunity was taken to showcase them, which did wonders for their pride in themselves and their unit.

And then there was learning to speak German. Somewhere up

the line a wise decision was made that we should have a rudimentary ability to speak German. The Army had a scripted curriculum consisting of various dialogues which we had to memorize. There were enough German wives in the battalion to instruct as well as provide practical information on getting along in Germany.

At the end of the course, we were in no position to engage in a philosophical discussion in German, but we could order food in a gasthaus, negotiate the train system, go shopping, give directions, tell time, count, and generally get along. I still remember the first scripted dialogue:

"Ist das der bahnhof?"
"Das ist nicht der bahnhof."
"Wo ist der bahnhof?"
"Gehen sie geradeaus, und den die zweite recht…" And so on.

There were other, smaller events, taking place. They were insignificant in the greater scheme of things, but of great importance to the principals involved. One such was the marriage of one of my soldiers. PFC Melvin Dale of 1st Platoon was engaged to be married to a local girl, and he invited me and his platoon leader, Lieutenant Ward Jones, to the wedding, which was to take place in the brigade chapel.

On the appointed day, Ward and I were seated in the chapel along with several of PFC Dale's fellow soldiers. While waiting for the ceremony to begin, PFC Dale came up to me and said, "Thanks for coming, sir."

"Don't mention it. I wouldn't have missed it."

"Uh, sir, would you mind giving the bride away?"

I was gob-smacked but managed to appear as if I received this kind of request routinely. "I'd be honored to do that."

"Thank you, sir."

"If I'm going to give her away, I suppose I should meet her first."

"Uh, yeah, I mean yes, sir. She's in the back."

PFC Dale escorted me back to the chapel entrance, where his bride-to-be was waiting in a side room just off the entrance. After the introductions, PFC Dale left to take his position at the altar while his fiancée and I made small talk. I thought she was a pretty tough nut and was glad it was his wedding and not mine. As the recording of the wedding march played over the sound system, I escorted her down the aisle with all the solemnity of a father and his daughter, handing her off at the altar to the man who would very shortly be her husband.

Such were the kinds of incidents that made us more like a family than a mere functional organization. Very soon, our very identity as a family was to come under threat.

The 1st Infantry Division is the oldest and most storied infantry division in the U.S. Army. As such, its lineage and honors were jealously guarded. The 28th Infantry Regiment was one of the original 1st Division regiments, the first to set foot in France during WWI and the first to go into battle, winning fame and glory for the division at the Battle of Cantigny.

As we prepared to move to Germany, it was deemed unacceptable that the Big Red One would lose its 28th Infantry battalion to another division, so the decision was made to redesignate the battalion as the 2d Battalion, 16th Infantry Regiment. This caused huge consternation in the battalion. The 16th Infantry is a fine regiment, second to none, but it wasn't us. We were the 28th Infantry, we were Black Lions, possessed of a high level of unit pride, and we didn't want to be anything else.

As we stewed at the prospect of having to become something

we weren't, someone with a knowledge of history and the Army's regimental system came up with the perfect solution. The 3d Battalion of the 28th Infantry had been deactivated in the 1950's and had laid dormant since then. Why not reactivate the 3d Battalion, redesignate us as such, and leave the 1st Battalion at Ft. Riley?

That way, the 1st Division could retain the 1st Battalion at Fort Riley, and we could remain Black Lions, albeit as the 3d Battalion instead of the 1st. There was initially resistance to this proposal on the part of some bureaucrats but, in the end, it was seen as a solution that made everyone happy and was adopted.

At a battalion ceremony, we relinquished the colors of the 1st Battalion, 28th Infantry and resurrected the colors of the 3rd. We were happy.

On the personal side of things, there were houses to be sold, arrangements to be made for the shipping of cars and household goods, passports acquired, and sundry other details necessary for getting families to Germany once the battalion was established there. All in all, it was a very busy time and, before we knew it, our deployment date of 12 November 1976 had arrived. Families would stay behind in the U.S. to await their turn to join us at a later date. We were packed and ready to go.

The out-processing center was set up in the gymnasium and we were to begin our embarkation at midnight. In the interim, we were on our own time. I had already come up with a plan to leave a calling card, so to speak, and had assembled the necessary supplies, which consisted of black spray paint and several different-sized stencils of our regimental insignia, which was a black rampant lion.

We now return to the beginning of this story, with the platoon leaders and platoon sergeants assembled in my office to receive orders for the night's mission.

"1st Platoon is to take the division museum area and change

the bumper markings of all vehicles. Under the old regimental system, we would be I Company, so paint the markings to reflect I-6, 3-28 IN and add black lions at appropriate places on the vehicles."

"2nd Platoon will take Custer Hill at large, giving priority to all lamp posts and telephone poles. Take advantage of targets of opportunity."

"3rd Platoon will take 1st Brigade headquarters. You will send a team to report to the staff duty officer as a "cleanup detail". Once inside, paint black lions wherever you can do so without being detected."

"I will be located in my rear command post, meaning I will be out of touch. Once the alarms go off, everyone will be trying to get hold of me to stop the operation. If they can't find me, they can't stop it. Are there any questions?"

Ever since I was a kid, I enjoyed pulling pranks. In the impulsiveness of youth, I never thought far enough ahead to worry about consequences. But now, in my mid-twenties and responsible for a company, consequences were something I had to consider, both for my men and for myself. In contemporary military parlance, this is called a risk assessment.

In an athletic contest, once the game starts and all the unknown factors enter, the coach is no longer in control of what happens to his team on the field of play. I had to consider what could go wrong with this operation. Having determined that there was minimal physical danger to the troops involved, I concluded that the greatest danger was that of getting caught, possibly arrested by the MP's.

However, since the "Opponents" would be unsuspecting, I felt that the risk was acceptable. Through the use of stealth and ingenuity, I was confident that my men could achieve surprise and pull this off without getting caught.

In thinking about the risk to myself, I decided that the risk level was acceptable. Since we had, in essence, already left the division, there was nothing official that could be done to me by the brigade or division commander. The main risk was how our battalion commander, Colonel Jaco, would take it. If he reacted really badly, he could take it out on me on my next efficiency report, effectively bringing my career to a halt.

My assessment was that this wouldn't happen. Colonel Jaco was not one to overreact or to hold grudges. Who knew, he might even see the humor in it.

As predicted, it wasn't long before the alarms went off and everyone was trying to find me, but I was nowhere to be found.

I was feeling a little sheepish when I showed up at the gymnasium for our out-processing, not knowing what the reaction had been or how I would be received. The first thing that happened was that Captain Jim Hosel, commander of Company C, came running up to me.

"Where the hell have you been, Buzz? The old man has had us looking all over for you. The whole post is in an uproar. Are you crazy?"

"Oh, you know, I just thought it would be nice to leave a little something to remember us by. I've just been hanging out in my BOQ room. There's no phone there."

Much to my relief, Colonel Jaco was surprisingly good-natured about the whole thing, as was Major-General Benedict, our commanding general. He approached one of my soldiers and said, in a joking manner, "I understand you like to paint black lions?"

"Yes, sir", the soldier replied, "we painted one on your car."

We hadn't, but I'm quite sure that General Benedict thoroughly checked out his car when he got back to it.

Once we entered the gym, we were in lock-down mode. There was no leaving of the building once we entered. This was for security reasons and concerns about drugs. We were not even allowed to go into the bathroom to relieve ourselves. This was quite a hardship because out-processing was long and drawn out.

For PFC Bischoff of my company, it was too much, and he peed his pants. His good fortune was that our plane landed in Bangor, Maine to refuel and it happened that Bischoff was from Bangor. Somehow we got a message to his parents, and they were able to bring him some clean underwear.

We left Fort Riley in the middle of the night, and with time changes, we managed to stay in darkness all the way to Frankfurt. As we approached the terminal after debarking the plane, two startling sights greeted me. The first was a pair of polizei on patrol. Both wore distinctive, Nazi-style visored caps and one of them had a sub-machinegun slung over his shoulder.

In 2024, this may not seem noteworthy, but in 1976, to see a policeman going so armed was menacing.

The second indelible memory was the bright orange plastic storefront of Dr. Muller's Sex Shop, which had on display in the window all sorts of sex toys and implements that would have been highly taboo in the U.S. (Later, we bought a life-size blowup sex doll, dressed her in our company Physical Training outfit, named her Black Lion Bertha, and presented her to Colonel Jaco. Unlike the rest of us, he had come over unaccompanied by his family. When his wife Carolyn came over for a visit, Colonel Jaco had to hurriedly stash Bertha out of sight. It was merely a joke gift, but it seemed prudent not to have to explain it).

Loaded onto buses and vans, we were taken from Frankfurt to Wiesbaden and our accommodations for the night. Driving through the streets of Wiesbaden in the dark, I was dazzled by the sight of half-timbered buildings, just as I had seen on picture postcards. Kansas was far behind us and a new adventure had begun.

Regimental insignia of the 28th Infantry "Black Lions of Cantigny"

18. Fulda

Fulda. The Fulda Gap. These are words that will be famil-iar to anyone who served in Germany during the Cold War. The Fulda Gap was one of the main avenues the Soviets were expected to use in the event of an attack on West Germany. Accordingly, the defense of the Fulda Gap was the primary concern of V Corps, of which we were to be a part.

It seemed like the Black Lions had no sooner arrived in Wiesbaden in November 1976 than we were being hustled up to our defense positions in Fulda. Major General Cleland, the commanding general of the 8th Infantry Division, which had operational control of our brigade, seemed convinced that war with the Soviets could break out at any moment and that we should not waste any time getting ready.

Responding to this sense of urgency, our commander, Lieutenant Colonel Tom Jaco, hastily ordered a reconnaissance to Fulda for himself and the company commanders. Our Operations Officer, Major Al Wissinger, planned the trip. He assumed we would be able to find overnight accommodations, either at a U.S. base or at a gast haus. But, just in case, we were to bring our bed rolls.

This turned out to be a sound precaution because there were no overnight accommodations to be had. Leaving us to scout the terrain, in spite of the fact that we didn't yet have a specific defense sector assigned, Major Wissinger went to find some place where we could spend the night.

"I'll be back as soon as I can. I'm going to go find us a place to stay."

Ever ready to grouse, I said, "Oh, this is great. They rush us up here without giving us an assignment and no place to stay. It's more of that old Army game, 'You play ball with me, and I'll shove the bat up your ass.'"

After looking around for a few hours, not really knowing what we were looking at, Major Wissinger caught up with us. "I found us a place to stay."

Major Wissinger was always positive and upbeat, so his pronouncement lifted our spirits. He was an old hand and very competent. We knew he'd come through for us. Following him in our jeeps, anticipating a warm bed and maybe even a beer at dinner time, he led us to … an abandoned barracks!

It was cold, dank, and filthy with trash everywhere and old, damp mattresses laying on the floor.

As we examined our digs, there were unpleasant discoveries. "Goddamn, there's a dead rat over here …'

"Jeez, someone took a shit on the floor."

"Well, look on the bright side, at least the metal bunk frames are here, and we won't have to sleep on the floor."

"This is really fucked up. They drag us up here like it was a crisis, without a mission and no place to stay."

Things improved dramatically after this inauspicious beginning. We were assigned a specific sector to defend and a gast haus was found that became our home away from home on our many trips to our GDP (general defense position), so much so that the owners housed our jeep drivers in their own residence while Colonel Jaco, Major Wissinger and we company commanders stayed in the gast haus. There were two of us to a room, each sharing the same bed.

Jim Hosel, the commander of Company C, and I shared one room, with Dennis Jacobson and John Murphy sharing another. The major and the colonel, being field grade officers, had their own room. As stated, the drivers stayed in the private residence, where they all fell in love with the owner's pretty, young daughter. That, perhaps, speaks volumes about the trust they had in us and our men.

The sector we were assigned to defend was on the eastern edge of the city of Fulda on high ground overlooking the north-south autobahn which was about one kilometer to our front. Because of the obstacle created by the autobahn embankment and the undesirability of attacking through a large urban area, an attacking Soviet force would have been obliged at this point to turn south and skirt around Fulda. My company position included what would have been the northwest corner of their turning movement, providing multiple flank shots at their tanks and armored personnel carriers.

I walked every inch of my position, selecting the exact location for every anti-tank weapon and attached tanks. This involved walking through the neighborhood and selecting specific houses into which our tanks would drive, in order to get a good shot from the concealment of the house.

Heavy TOW anti-tank weapons were located along the autobahn where the embankment would provide protection from direct fire while providing good fields of fire. The DRAGON medium anti-tank weapons were organized into tank-killer teams east of the autobahn where they could achieve flank shots from the concealment of a railroad embankment. My mortars were placed in a school yard on the reverse slope of the high ground we were on. Under the tutelage of Colonel Jaco and Major Wissinger, we collectively went over every detail of the plan until we knew it backwards and forwards.

After scouting our positions all day, we would return in the evening for a savory meal of jaeger schnitzel washed down with beer and complemented by congenial conversation. Somewhat rare in the Army, Colonel Jaco and Major Wissinger were both highly professional and affable, making personal interaction easy and relaxed. We would have gone anywhere and done anything for Colonel Jaco.

The commander of V Corps at the time was Lieutenant General Donn A. Starry, the author of the concept of the air-land battle whereby an outnumbered U.S. force would successively defend against a Soviet invasion of West Germany. He took the business of war fighting very seriously and expected no less from his subordinates.

Putting his time and energy where his mouth was, he required every commander in his corps, down to company command level, to personally brief him on their plan to defend their sector of the Fulda Gap. These briefings took place on the ground at each subordinate commander's defense position. In V Corps, this involved briefings by approximately one hundred company commanders, not counting the briefings by battalion and brigade commanders.

It was rumored that General Starry would relieve on the spot any commander whose briefing was unsatisfactory. No one wanted to be the one to make the rumor a reality.

In the military, commanders are responsible for everything their subordinates do or fail to do. Consequently, company commanders had to brief everyone in their chain of command before briefing the corps commander: battalion commander, brigade commander, assistant division commander, division commander and, finally, the legendary General Starry himself.

Because we had painstakingly developed the plan in the battalion, briefing the battalion commander was easy, but also a good dress rehearsal for the briefings further up the chain of command. Briefing the brigade commander and the assistant division commander went smoothly but briefing the division commander was going to be a challenge.

Major General John R.D. Cleland was a small man with wire rim glasses who looked every bit like a kindly old grandfather. But looks can be deceiving. General Cleland was a wild man who had commanded troops in combat in World War II, Korea, and Vietnam. With the attachment of our brigade to his 8th Infantry Division, he commanded the largest division in the U.S. Army, which pleased him no end.

As the result of a battle wound, one of his hands was atrophied into a permanent 'thumbs up' gesture with which he greeted everyone while yelling "Combat Ready!" He seemed to believe that the war with the Soviets could happen at any moment, and he commanded his division accordingly.

He even directed all his units to be on their battle positions on Christmas Eve, although he subsequently relented on that. Briefing him would be no piece of cake.

At the appointed day and time, we were set up according to the format prescribed by General Starry. The site would be a position from which the whole sector could be seen. The briefing would take place on the hood of the company commander's jeep, using a map and the battle book which contained an overlay for every conceivable aspect of the battle: the direct fire plan, the indirect fire plan, the obstacle plan, the plan for utilizing Army and Air Force aircraft, and on and on.

Everything was going smoothly with General Cleland until he said, "Why did you put your DRAGON teams there?"

At 28, I still hadn't outgrown the chip that I'd carried on my shoulder since childhood, and any perceived slight could ignite my anger. There was something in the way General Cleland asked the question that pushed my hot button.

"Because it's a damn good position, that's why!" I retorted in an angry voice.

Colonel Jaco was quick to react, verbally leaping between General Cleland and me. "What Captain Sherwood is trying to say is that this position places the DRAGONS within the range at which they are most effective, provides flank shots, and utilizes the cover and concealment of the railroad embankment."

To my relief, General Cleland gave no indication that he noted my inappropriate response, and he seemed satisfied with Colonel Jaco's interjection.

The remainder of the briefing went without mishap.

That evening, we had a laugh over my close call. Other commanders might have seen it as cause for a severe ass chewing, but that wasn't Colonel Jaco's style.

Finally, the day came to brief the big guy. Having poured our professional talent into developing a sound plan and testing it against everyone up through the division commander, we were confident that we would pass muster with General Starry. And yet, and yet, ... who knew what could go wrong? We weren't home free until we were home free.

I was at my briefing position on the Rauschenberg, a hill on the eastern edge of Fulda, waiting my turn when I got a call on the radio from Colonel Jaco.

"Alpha-six, this is Black Lion-six. We are finishing up here, en route your location, over."

"This is Alpha-six. Roger."

Pretty soon, a small convoy of jeeps approached, radio antennae swaying. Leading the way was Colonel Jaco. Dismounting, he gave me a surreptitious wink as he assumed his position to the rear of where General Starry would stand.

Next came General Starry, followed by a bevy of staff officers and other strap hangers. Like General Cleland, he looked nothing like his fearsome reputation led me to expect. He wore glasses and had a fleshy face, more like what you'd expect a professor to look like rather than a daring tank commander.

What was really striking was his outfit. He was wearing one piece tanker's coveralls and a World War II style tanker's jacket. On his feet he was wearing galoshes, but not the GI kind with buckles. These had a zipper up the front and the toes were pointy, as if he were wearing them over cowboy boots. Who knew? It was fairly common for generals to take liberties with their uniforms.

"Good morning, sir! Captain Sherwood, Alpha Company, 3d Battalion, 28th Infantry."

After he returned my greeting, I proceeded with the briefing according to the prescribed format. "Sir, we are currently located at this spot on the eastern side of the Rauschenberg."

"What is the elevation?"

"Sir, the elevation is four hundred fifty-five meters."

With that, he reached in his pocket and pulled out some calipers, bent over the map, and measured the contour scale.

"Where do you expect to start servicing targets with direct fire?" (I was amused by his use of the term 'servicing targets', conjuring up as it did the image of a bull being turned loose in a pasture full of ripe cows).

"Sir, we'll engage targets when they hit the maximum effective range of our TOW's, which is shown here on the map."

Out came the protractor from a pocket to measure the distance from my TOW positions to the maximum effective range of 3,000 meters.

"Where is that on the ground?"

As I began pointing out identifiable terrain features, he whipped out a pair of binoculars. It seemed he had a gadget in his pocket to check the accuracy of every answer.

"Where is your air release point?"

Pointing to a spot in the sky above us, I said, "Sir, it's right there."

There seemed to be no gadget to check that, but this was a question we anticipated and were ready for.

The briefing proceeded with me explaining every minute detail of the battle plan and General Starry checking the accuracy of my responses from his trove of tools.

Finally, the briefing came to an end. The absence of any criticism was acknowledgement enough that I had met his high standards. Nevertheless, as the mechanized caterpillar drove away with its multiple antennae swaying back and forth, I heaved a great sigh of relief.

That evening in the gast haus, we had ample cause to celebrate. We had all passed a rigorous test. Each company commander could take pride in their own achievement, but we all knew we wouldn't have gotten there without the guidance of Major Wissinger and Colonel Jaco.

Over dinner, Colonel Jaco raised his beer stein, "We did good today."

"Black Lions, sir" came the reply.

A Company, 28th Infantry, Wiesbaden, Germany

19. Who's Who Here?

In June 1977, the Black Lions entrained for Wildflecken to do squad and platoon level training. Wildflecken was a training center located in a hilly, wooded and remote part of Bavaria. During World War II, it was a training camp for the Wehrmacht and the Waffen SS. At the main gate was the remnants of a concrete Nazi eagle. It was headless and wingless, and the Nazi swastika was gouged out of the wreath held in the eagle's talons.

One evening during a lull in the training I decided to take my officers out for a quiet evening at a gasthaus. My Executive Officer, Lieutenant Willie C. "WC" Garrison, and platoon leaders, Lieutenants Bill Lockwood and Ward Jones, piled into my jeep along with my driver and ever-present companion, Specialist 4th Class Kenny Hall.

We headed into a small village off the Wildflecken Kaserne. The American presence in Germany was so pervasive that the appearance of GI's in a gasthaus was unremarkable. Many a time, Kenny Hall and I would park the jeep in front of a gasthaus, walk in and take a seat, hanging our rifles on one corner of the chair, our helmets on the other, without any reaction whatsoever from the staff or customers.

But this time it was different. It appeared that the village we selected was not used to the presence of GI's, judging by the sensation we created as we walked in, with heads turned and faces reflecting looks of surprise.

As we came in, I noted a composite picture on the wall near the door showing the citizens of the village who had served in World War II. Some wore the uniform of sailors, but most had been soldiers. Most of the customers seemed to be of that vintage. It quickly became apparent that no one spoke English. Fortunately, we spoke enough German to do the basics like ordering food and drink. After seating ourselves at a large table on the outer wall, I ordered for the group.

"Ich mochte ein grosse bier fur alles, bitte."

Soon we sat enjoying large steins of good German beer. As we sat chatting, a few of the old gentlemen wandered over to our table.

'Hmmmm, I wonder what they want. Are they going to fight the war all over again? Are we unwelcome intruders here?'

"Ich war soldat."

His companions nodded as if to affirm that they, too, had been soldiers.

In spite of the language barrier, it quickly became apparent that they were approaching us not as former enemies, but as men who shared with us the universal experience of soldiering and who felt a bond with us that spanned generations and nationalities. It was kind of touching.

After these gestures of bon homie, one of the gentlemen indicated that he wanted to know who we were. I knew German well enough to identify everyone by name and rank, but I wanted to give them more and I didn't have enough command of the language to do that. What to do?

Then I came up with an idea. The Germans at that time loved American cowboy culture. At Fasching, dressing as cowboys and

Indians was very popular, as were American movies and television programs with a western theme.

"Verstehen sie 'Bonanza'?"

"Ja, ja! Bonanza! Bonanza ist gut."

"Zehr gut. OK. Ich bin Pa."

"Oh ja."

WC Garrison, my XO, was clearly the counterpart of Adam Cartwright, the oldest son on Bonanza who kept his younger brothers in check. Gesturing toward WC, I said, "Das ist Adam."

"Oh ja, ja. Zehr gut."

Bill Lockwood, my light-hearted, sometimes silly platoon leader seemed the perfect match for Little Joe. Indicating Bill, I said, "Das ist Little Joe."

More nods of understanding and approval from the Germans.

Next, I introduced the serious, somewhat dour Ward Jones. There being no more Bonanza sons to choose from, Ward was obviously going to be Hoss. It wasn't a bad match except that Ward wasn't built like a linebacker, as was Hoss. Again, great approval of this introduction.

Feeling very pleased with myself, my attention settled on my driver, Kenny Hall. I had to introduce him as well, but which Bonanza character could I match him with? Wanting to complete the introductions and not wanting to exclude Kenny, I was running the Bonanza cast through my mind, looking for the right match. Once again, the lights went on. "Hop Sing!! Das ist Hop Sing."

This produced gales of laughter all around along with expressions of approval. Kenny Hall grinned broadly, his white teeth gleaming against his dark skin.

Pretty soon, the language barrier prevented further interaction, and the Germans drifted back to their places. Not long after, we finished our beers and made our way to the door amidst friendly farewells. We left with a good taste in our mouths and hoped that we left the same.

* * * * *

Towards the end of June we got ready to re-deploy to Wiesbaden. Re-deployment from the field in Germany was always a major production. Barracks had to be cleaned and inspected; cots, stoves and other equipment cleaned and turned in; vehicles washed and maintained, and all of this on a tight schedule that accommodated all the other units that were re-deploying at the same time.

For this move, our tracks were to be transported by lowboys while the troops would travel by train.

By this time, we had been on numerous deployments and were used to the drill, going through the paces very matter of factly.

At the appointed time, we assembled at the railhead to board our train. Another unit's train was on station, its last car sitting right in front of us, but no sign of our train. Yet another instance of "hurry up and wait." After waiting for what seemed like hours, one of my soldiers approached me. "Sir, I gotta take a crap."

"Isn't there a latrine here?"

"No, sir."

"Well, can't you wait?"

"No, sir, I gotta go bad."

"All right. Jump on that train there and use the toilet, but hurry up, it's going to leave pretty soon."

Off he went like a shot, climbing up the steps and entering the compartment. Minutes seemed like hours as I watched anxiously for him to reappear. "What the fuck is he doin' in there, readin' a book?"

All of a sudden, the train started making the puffing and clanging noises that signaled it was getting ready to depart. Now I was really concerned at the prospect of one of my men getting sepa-rated from us and transported to God knew where. I stared at the back of the car, willing him to appear.

Just as the train started to move, he appeared at the rear of the car, pulling his pants up as he exited the car and leaped off the small platform at the rear of the car. Everyone on the platform was cheering wildly.

Thus ended our deployment to Wildflecken. All's well that ends well.

20. Hindut Ninyo

In November 1977, I relinquished command of Co A, 3rd Battalion, 28th Infantry in Wiesbaden and immediately assumed command of Headquarters and Headquarters Company (HHC), 8th Infantry Division (8th ID) in Bad Kreuznach. Shortly before my move, Major General (MG) Paul F. Gorman had taken command of the 8th Infantry Division from MG John R.D. Cleland.

A couple of months into my command of HHC, I was notified that I had been selected to have dinner with MG Gorman and LTG Sidney Berry, the V Corps Commander, along with a couple of other company commanders from the 8th ID. I don't know why I was selected. Possibly because I had successfully commanded a rifle company, possibly because I had commanded a company in the 1st Infantry Division, the Big Red One, in which MG Gorman had served in Vietnam, and possibly because I was conveniently located at HQ, 8th ID.

In any case, I found myself at the appointed time at the designated German restaurant with the two generals and other captains who had been invited. As the waitress made the rounds, taking orders, General Gorman, who was fluent in German, made his selection. General Berry, who appeared not to speak German, pointed to an item on the menu. I spoke enough German to get by, but not enough to peruse the menu for some exotic gastronomic extravaganza. I ordered Jaegerschnitzel, a tasty dish, and usually my default selection.

A few minutes after ordering, the waitress came back and questioned General Gorman about General Berry's meal selection. It turned out that he had ordered shark, and they wanted to make sure that was what he wanted. Appearing to be both embarrassed and amused, General Berry changed his order to something more in the line of meat and potatoes.

This was not strictly a social outing. The purpose of the get together was to solicit input from successful company commanders about the current state of company command and what needed to be done to improve the quality of company commanders.

In 1977, the Army was still in the process of building itself back from the turmoil of the Vietnam war. Much progress had been made, but there was still a long way to go.

I gave my honest assessment and frankly answered their questions. Of those exchanges, the only one I remember is General Gorman asking me if I thought we should have majors commanding companies, as was the case in the British Army. I stridently rejected that notion, stating that I believed that captains were quite capable of successfully commanding a company.

General Gorman was sitting directly opposite me and, with a twinkle in his eye, he said to me, "Do you know the motto of the 1st Infantry Division?"

"Yes sir! 'No Mission Too Difficult, No Sacrifice Too Great. Duty First!'"

"No, no. That's the sissy one."

With the sparkle in his eye and the corner of his mouth turning up in a sly smile, he said, "It's 'Fuck 'em All!'"

As a history buff, I was surprised that I had somehow missed this nugget, but I liked it. I liked it a lot. I immediately suspected that this motto dated to WWII when the Big Red One was commanded by the legendary and beloved MG Terry De la Mesa Allen, who cared only about his unit and his troops, damn the rest. This sentiment appealed to me greatly.

Back at the company, I resolved that this would become our company motto. I wasn't totally stupid, though, and I knew we couldn't blatantly adopt this in plain English. We would have to be discreet. There were two Korean soldiers in the company, Kim and Park, and I landed on the idea of translating our new motto into Korean.

"Get me Kim or Park," I called out to the First Sergeant.

Kim was one of the company clerks and was normally readily at hand, but for some reason neither he nor Park seemed to be around.

Excited by my grand idea, I was waiting impatiently for Kim or Park to be found when, who should walk by my office but Soriano, a member of the company who was Filipino.

"Soriano!" I called out.

Momentarily, Soriano reappeared in my doorway.

"Yes, sir?"

"Come on in. Do you speak Tagalog?"

"Yes, sir."

"Outstanding. I need you to translate something. How do you say, 'Fuck them all' in Tagalog?"

"That would be 'Hindut Ninyo'"

"How do you spell that?"

"H-I-N-D-U-T N-I-N-Y-O"

"Excellent; thank you."

And so, HINDUT NINYO became the unofficial moto of HHC, 8th Infantry Division. It was painted on the large company sign in front of our barrack and was printed on our company T-shirts and sweatshirts. We shouted it as a quasi-battle cry in company formation, and we chanted it as we ran in formation around Rose Barracks during morning physical training.

For many months, it was our little secret, hidden in plain sight. Hidden, that is, until I received a call from Lieutenant Colonel Jerry Hogan, the Inspector General for 8th ID. "Sherwood, what the hell are you doing down there?"

"What do you mean, sir."

"I just got a call from a dependent who was walking through the kaserne and saw your sign which, she said, has 'fuck them all' painted on it."

"Oh, that," I chuckled. I then proceeded to relate to Colonel Hogan the story of my dinner with General Gorman and the process of arriving at Hindut Ninyo.

In a surprisingly tolerant response, Colonel Hogan said, "Well, get it off the company sign. I don't want any more complaints from dependents."

"Roger that, sir."

The motto was dutifully painted over on the company sign but all else remained the same throughout my tenure as commander. I don't know if it survived after my departure, but it is remembered by those who served in the company at the time. At reunions, someone invariably brings it up.

General Gorman, busy commanding the largest division in Europe, probably never knew the ripple effect of his casual dinner table comment. Hopefully, it would have given him a chuckle.

HHC, 8th Infantry Division, Bad Kreuznach, Germany

21. Lead Me Home

In November 1977, I completed my command of a rifle com-pany in Wiesbaden, Germany and was transferred to Headquarters, 8th Infantry Division at Rose Barracks in Bad Kreuznach where I took command of the division Headquarters Company.

Every morning before work call, we did physical training (PT). Occasionally this involved exotic activities like sports, Rifle PT, or bayonet training, but typically it consisted of calisthenics and a two-mile run. Normally we ran in formation twice around the kaserne.

One day in July 1979, I decided to go on a longer run. When we got to the back of the kaserne by the 8th Signal Battalion, we went out the back gate, turning right onto the road separating the base from the German community. Not long after making the turn, I noticed from my position at the front of the column that someone had fallen out of the run. This wasn't unusual and could be due to a sprained ankle, fatigue, untied bootlace, or any number of things. What caused me some concern was that this individual not only fell out, he fell down. I further noted that one of the medics fell out of formation to tend to him. Satisfied everything was in hand, we continued on the run.

One of the unique aspects of military life is that the mission takes priority over everything else. The human instinct is to render assistance, but in the military this instinct is replaced by an imperative to keep moving forward to accomplish whatever the mission is at that particular moment. There are those whose job it is to render assistance. For the rest, the mission continues.

A few minutes later, we heard the sound of sirens. I had the ominous feeling the sirens were connected to the soldier who fell out of the run. I wanted to be instantly transported back to the barracks to find out the situation, but the reality was the fastest way back was to continue running.

Upon arriving back at our barracks, I dismissed the company and ran into the orderly room. "What happened?"

"It's Sergeant West, sir, he died."

"HE WHAT?"

"He died, sir. It must have been a heart attack."

"Where is he?"

"Up at the dispensary, sir."

"Get my jeep, quick."

Specialist Bertrand appeared shortly with my jeep, and we raced up to the dispensary. Not being a large facility, it didn't take long to find where we needed to go. The door to the examining room was open and Sergeant West was laid out on the examining table with the Division Surgeon, Colonel Bzik, and some medics in attendance.

I sat down on a chair outside the room, feeling miserable. For no rational reason, I was blaming myself and all kinds of thoughts were going through my mind.

'If only we hadn't left the kaserne, if only ... what if ...' and on and on.

After a while, Colonel Bzik came out. Drafted into the Nazi army in World War II and later commissioned in the U.S. Army as a doctor, he was a hard man, not much given to sentimentality. Any love he had to dispense was of the tough variety.

Jumping to my feet to face him, he immediately read me like a book.

"Zis is not your fault," he barked at me as if it were a reprimand, "Stop feeling sorry for yourself. He vas dead before he hit the ground. Zere is nothing zat you or anyone else could have done."

"Yes. Sir." I muttered, wondering if I was that transparent. I also immediately realized his approach was the right one and had effectively snapped me out of self-blame and self-pity. I was the company commander and there was work to be done.

Mortuary Services would take care of shipping the remains back to the States, but there was much I had to do. In addition to seeing to administrative and financial details, Sergeant West's personal belongings had to be packed up and a memorial service organized. Master Sergeant Scott would take care of the administrative and financial part, Sergeant West's assistant, Sergeant Matlock, would take care of the personal effects, while I met with the chaplain to plan the memorial service, which would be held in the chapel.

Military memorial services are pretty cut and dried with little left to the imagination. I would give the eulogy and help select the readings. Chaplain Alexander, a Major, and I worked out the readings and set a date.

A couple of days later, First Sergeant Schenkenberger poked his head in my office.

"Sir, PFC Jones would like to speak to you."

"OK, send her in."

A moment later there was a knock on my door.

"Come in."

A tiny African American soldier entered and marched up to my desk. Soaking wet and in her combat boots, she couldn't have weighed more than ninety pounds. She saluted and said, "Sir, PFC Jones."

Returning her salute, I said, "At ease. What can I do for you?"

In a tiny voice she said, "I want to sing at Sergeant West's service."

Touched by her interest in doing something special and knowing the memorial would be a family type affair, I didn't feel the need to hold an audition or otherwise determine her qualifications. I went back to see the chaplain.

"Chaplain, one of the soldiers in the company would like to sing at Sergeant West's memorial."

Expecting this to be nothing more than a courtesy announcement to the chaplain, I was taken aback by his response.

He said, "We don't normally allow that," as if no more needed to be said.

His officious attitude pushed a hot button in me and now it was the chaplain's turn to be taken aback.

"Well it's not your memorial service is it? Sergeant West is the one that's dead and PFC Jones would like to sing at his service."

Backing down, he said, "I guess we could allow it."

At the memorial service, after the readings, sermon, and comments by myself, it was time for PFC Jones to sing. I knew her gift would be beautiful, simply because it was her gift, but I was expecting to hear something like a baby bird cheeping out a little song.

PFC Jones stepped up to the pulpit, opened her mouth and began to sing.

"Precious Lord, take my hand, lead me on, let me stand…"

What? This was not the voice of a baby bird. This was a

full-throated gospel singing voice. It was as if this tiny person was inhabited by Mahalia Jackson or Aretha Franklin.

"Through the storm, through the night, lead me on to the light, take my hand precious Lord, lead me home.

As she reached a crescendo of controlled screaming, the power and beauty and sheer rawness of it gave me goose bumps.

"Hear my cry, hear my call, hold my hand lest I fall."

The hymn reached its climax, and all sat in stunned silence with tears flowing and mouths agape.

"And the day is past and gone
At the river I stand
Guide my feet
Hold my hand
Take me home"

There was nothing more to be said. Nothing more to be done. Sergeant West was safely across the river and home.

22. For Your Action As You Deem Appropriate

In Headquarters Company, 8th Infantry Division, when I commanded it, there were two Koreans who were members of the company. Their names were Park and Kim. Park spoke English well, but Kim spoke very little. Of all assignments, Kim was a clerk in the orderly room. He may not have understood much of what he typed, but he was able to type to perfection whatever was put in front of him.

Both Park and Kim were very good soldiers, but Kim's inability to speak English was an obvious impediment to his advancement. His supervisors and I were always looking for ways to help him improve his language skills. It occurred to me that Park would be the obvious one to help him, so I called Park into my office.

"Park, you know Kim's English isn't too good. What do you think about helping him to learn?"

"I can't do that, sir."

Helping a fellow soldier, especially one of the same national origin and primary language, seemed an obvious solution and I was quite taken aback by Park's refusal.

"What do you mean, you can't?"

"Sir, he's older than me."

"Well so what? What does that have to do with it."

"Sir, if I was to do that it would cause him to lose face. I can't do that."

I knew about the importance of saving face and losing face and that I had to respect Park's position.

"OK, I get it. But if there's any opportunity to help him without causing him to lose face, do whatever you can, OK?"

So there matters stood for the time being. Meanwhile, other events were taking place in Kim's world.

The company had a mascot dog, a German Shepherd-Doberman mix named Aeris. She was a sweet dog and hung around the orderly room most of the time. Aside from his clerical duties, it was Kim's duty to feed her. Every day he would go to the supply room, get a plastic bag and then go to the mess hall for food scraps.

The supply clerk who doled out the plastic bags was Private DeMott, who hailed from New York City. He was arguably the most scrawny, unwashed, unskilled soldier in the company. He couldn't tie a necktie or even a bow knot in his bootlaces and required close supervision from Sergeant First Class Marsland, the Supply Sergeant.

Human nature being what it is, those with the least power will often exploit any opportunity to exercise power over others. Kim's daily visit to the supply room gave De Mott just such an opportunity. He was the keeper of the plastic bags. He was smart enough to know that he couldn't refuse Kim a bag, but what he could do was make the experience as unpleasant as possible.

Every day, in addition to collecting a bag, Kim also collected what in the army would be termed a "ration of shit" from Private DeMott. This caused Kim to lose face in a much more powerful way than would an offer of help from Park, but Kim was too good a soldier to complain. This daily torture went on for an undetermined amount of time because the only people who knew it was happening were Kim and DeMott.

Finally, Kim had taken all the abuse he could stand. One day I received in my in-box an official, typed memo from Specialist Kim. It had been routed through both the First Sergeant and the Executive Officer and both had put their "chop" on it, acknowledging that they had seen it and were forwarding it on to me for action. The subject of the memo was "Complain about the Supply Clerk PV2 Demott."

The gist of the memo was that Kim had taken all the abuse he was going to take, as indicated by the following excerpts: "Every time PVT Demott give to me hard time when I pick up some Plastic Bag (sic) for take the trash and pick up some dog food at the Mess Hall."

"I have so many time this same situation. Somewhen I will beat him if he get chang (sic) the mind, becaus (sic) I can't endure anymore. I am very serious because I am not beggar."

Subjected to this kind of abuse, most soldiers would have exploded in anger and punched the offender, but Kim was far too professional for such lack of self-control.

Knowing that the first sergeant and XO would have already looked into the matter, I called them both into my office. "What the hell is going on."

"Well, you know that Kim goes to the supply room every day to get a garbage bag for Aeris' food, and it seems that DeMott gives him shit every time he goes down there."

"How do you think I ought to handle it?"

We talked about the many different options. I could let the first sergeant handle it, since it was an enlisted matter well within his purview. I could call DeMott onto the carpet and order him to stop his harassment of Kim. I could call SFC Marsland in and put the responsibility for settling it on him, since De Mott was his clerk.

All of these were perfectly good solutions, and all would have produced the desired result, but they all had the same flaws, inasmuch as Kim would have lost face by having someone else solve his problem for him and DeMott would have the solution imposed on him from above, rather than doing the right thing of his own volition.

As a soldier, I wanted to be told what the mission was and then be left alone to accomplish it my own way rather than being told how to do it. I sought always to afford my subordinates the same courtesy. It occurred to me that I could apply the same philosophy in this situation.

"Let's do this: route this through SFC Marsland to DeMott with the indorsement, 'For your information and action as you see fit', and put it over my signature. That way, DeMott can decide for himself to quit fucking with Kim or get his ass kicked. Let him decide."

The first sergeant and XO agreed with this course of action and so it was done. Kim never had another problem when picking up the plastic bags.

23. Uncertain Battle

Bad Kreuznach, Germany, Spring, 1979, 6:00 a.m. I was sitting in my car listening to an audio cassette of the Scots Guards pipe band playing the '10th Battalion, Highland Light Infantry Crossing the Rhine'. I was listening to the relentless skirling of the bagpipes for the same reason that Highland soldiers did—to screw up my courage for impending battle. I was not preparing for battle against the Soviets—that was something I was trained for, was ready for, and even occasionally hoped for. No, this was a kind of battle for which I was totally unprepared. As a captain I was contemplating the prospect of saying 'no' to the Commanding General of the 8th Infantry Division and his Chief of Staff. Defying my superiors was the antithesis of my training and professional inclinations. To say I was nervous and uncertain would be an understatement.

Some months previous, the 8th Infantry Division commander, Major General Paul Gorman, established a new staff section in his headquarters, the express purpose of which was to manage resources and hold subordinate commanders fiscally accountable for the resources entrusted to them. It was aptly named the Resource Management Office (RMO) and was headed up by a lieutenant colonel.

The division Chief of Staff, Colonel Robert "Buffalo Bob"

Wagner, had direct oversight over this operation, as he did for the rest of the division staff. Each subordinate commander, mostly colonels and lieutenant colonels, submitted data on their annual needs for things like ammunition, petroleum products, repair parts, rations, etc. basically every consumable item needed to keep a unit operational. They were then allocated funds sufficient to meet their stated needs. Commanders were then responsible for budgeting their annual allocation of funds to cover all their requirements.

It was made clear that failure to do so would be reflected in one's annual efficiency report. To ensure that there was no misunderstanding, each subordinate commander was required to report to the RMO office and personally sign their budget in the presence of the chief of RMO, thereby acknowledging their responsibility for it. As commander of the 8th Infantry Division Headquarters Company, a separate company, I had my own budget and the same obligation.

I had no problem with this requirement. In fact, I applauded it. My executive officer, Lieutenant Fred Parker, was extremely competent. He had a very accurate knowledge of what we spent and when we spent it. Between he and Sergeant First Class Marsland, my supply sergeant, an annual budget was developed, by quarter, reflecting all the peaks and valleys of expenditures. I looked forward to committing myself to my budget and living within it. Unfortunately, it wasn't as simple as that.

One day, about a week prior to the great signing event, Lieutenant Parker came into my office, quite agitated.

"What's up Fred?"

"Sir, we have a problem!"

"What is it?"

"Division is screwing us on our budget."

"Whadda you mean?"

"The amount they gave us only covers what the company spends, but it doesn't account for all the stuff the CG buys. You know how he's always buying his training gizmos and gadgets, well, it comes out of our budget. You never know when he's going to buy something, or how much it's going to cost. There's no way we can manage a budget that way."[6]

"Have you talked to anyone at RMO about this?"

"Yes, sir. I talked to Captain Rausey about it. He's the one who is putting all this together."

"And what did he say?"

"He said he wasn't going to change it. He said not to worry about it. If we need more money, they'll give it to us."

"Well how the hell are you supposed to manage a budget like that?"

"That's the whole point, sir. You can't. We'll be constantly jumping through hoops trying to readjust the budget."

"So what do you think I should do?"

"I don't think you should sign it."

The logic of Fred's recommendation was inescapable. However, logic isn't the only thing to take into consideration in any institution. The old adage, 'go along to get along' certainly applied here. I knew that division would funnel me the money to cover the CG's expenses, even though it would make a hash of my book-keeping. If I refused to sign the budget, I would more than likely not be seen as a team player, which could be reflected

---

6. Major general Gorman's area of personal interest was in training innovations. He was constantly buying high end simulators and other items to test out his training ideas. He often purchased these on the German economy in what was known as a Local Purchase.

on my efficiency report and subsequently affect my chances for promotion.

The easy way, the safe way was to play the game. But if I did that, I would be betraying my subordinates and all the work they had done to develop and live within a budget. And, to play the game would fly in the face of the whole purpose of this exercise, which was to make commanders fiscally accountable. What to do? What to do?

So there I sat in my car, letting the bagpipe music wash over me and enter my blood as I readied myself for my morning appointment at the RMO to sign a binding document that wasn't worth the paper it was written on.

A few minutes later, at the appointed time, I reported in to the RMO office where the colonel who was the primary staff officer and his assistant, Captain Rausey, awaited me. The colonel greeted me. "Good morning, Captain Sherwood. Are you ready to sign your budget?"

The atmosphere was relaxed, since this had become pretty much a pro forma exercise.

"No, sir, I'm not ready. I'm not going to sign it."

Stunned silence ensued. The relaxed, friendly looks transformed into looks of incredulity with eyes wide and mouths agape.

I was now committed but still feeling very unsure of myself. I had tread into unfamiliar territory and chosen to do battle with unfamiliar foes. I was sure I was doing the right thing, but I wasn't so sure it was the smart thing. How was this going to turn out?

"What? Why not?"

I proceeded to explain about the CG's expenditures and that I couldn't possibly be expected to adhere to a budget with variables like that. This appeared to be something that the colonel was unaware of, as evidenced by his seeming loss for words.

Captain Rausey was not unaware since it was to him that Lieutenant Parker had brought the issue previously. Nor was he at a loss for words. "It's no problem. We'll just feed you more money when you run out."

"That's crap. You're asking me to make a commitment that we all know I can't keep. I'm more than willing to be responsible for that which I have control over, but not this."

Captain Rausey gave me the stink eye. If looks could kill, I would have dropped dead on the spot.

The colonel realized that we were at a deadlock. He couldn't force me to sign and it's very possible he saw validity in my position.

"Let me look into this and we'll get back to you."

As I left the RMO office, I could have sworn that Captain Rausey snarled at me.

I few days later, I was informed that the RMO would establish a separate account for General Gorman's purchases, managed directly by the RMO. I would be responsible only for what we as a company expended. We had won. I happily signed my budget.

This was a valuable experience. My very talented Executive Officer and Supply Sergeant identified an issue before it became a problem, but it fell to me as the commander to go up against the big guys in a fight that was way outside my comfort zone. I owed it to my subordinates to do so and, ultimately, I owed it to my superiors to insist that the system they had devised function as intended.

24. What's in a word?

Like most professions, the Army has its own unique vocab-ulary, some of which is necessitated by the technical aspects of the profession. Other words are the product of long tradition and serve to make the military profession separate and distinct from the civilian world. Examples of the former would be the plethora of alpha-numeric nomenclatures and abbreviations with which military conversations are liberally seasoned: M-16's, M-14's, M-60's, SAW's, LD's, LZ's, DZ's, FEBA's, etc... terms readily understood by those in the know, but mind boggling to the uninitiated civilian stumbling into a conversation or trying to read a military tome which is not user-friendly to a general reading audience.

Examples of the latter category are more generally understood, but not always. Soldiers don't eat in a dining hall, they eat in a mess hall; they live in a barrack, not a dormitory; they are confined in a stockade, not a jail, and so on.

This separate language is not a problem until the military world collides with the civilian world and then there can be confusion and misunderstanding, and, for some, humor.

Upon my return in 1979 from three years in Germany, I was assigned as the Operations Officer of the recruiting battalion in Indianapolis, Indiana. My responsibility was to oversee the recruiting for the Regular Army throughout the state of Indiana.

I monitored quotas, market shares, trends, changes to eligibility requirements, etc. and reported those findings to the commander.

My counterpart in Army Reserve operations was Captain Pam Delabar, who performed the same functions as me in recruiting for the reserves. She was a full time reservist, which, to my mind, meant that she was a civilian who wore a uniform to work every day. Unlike me she was a resident of Indianapolis, whereas I would be transferred somewhere else as soon as my two-year assignment was over. All of her friends were civilians, whereas I had no civilian friends. Her civilian friends frequently called her at work to chat. One day I answered the phone as follows: "Headquarters, Indianapolis Recruiting Battalion, Captain Sherwood here."

"Can I speak to Pam?"

"She's out of the net right now, but her blouse is hanging on her chair so she can't be too far. Can I take a message?"

In the pregnant silence that ensued, I realized how my response must have been interpreted but, rather than clarifying the matter, I decided to let it be. Why shouldn't I have a good laugh imagining what Pam's civilian friend was imagining? After all, I had conveyed the essential piece of information: Pam was not available.

(In the Army, the top part of the Class-A uniform is not a jacket or a coat, it's a blouse. Underneath the blouse is worn a shirt with tie. The blouse can be removed for comfort as long as one is indoors.

When several stations are tuned in to the same radio frequency in order to communicate with one another, it's called a net. When one goes off-line, they are "out of the net". Being "out of the net" is a common slang expression for being briefly away from ones normal place of duty).

25. Would You Like to be a Millionaire?

The important lessons of life often come to us in very un-expected ways. This is the story of one such lesson.

I grew up in modest circumstances in the East Bay of the San Francisco Bay Area. My father owned a music store in Berkeley and my mother worked at Herrick Hospital in the same city. We were what are nowadays called flatlanders, but when I was growing up, you were either a duck or a goat. The ducks lived on the flatlands where, it was said, they had to swim when the tide was in. The goats lived in the hills in the bigger houses and had the fancier cars. Being called a duck or a goat was spoken in jest, but within those labels was an acknowledgement of affluence or lack thereof.

The presence or absence of money was never a major topic of discussion with my parents. My brother and I always had enough, but there were few frills. There was no cabin at Lake Tahoe or Russian River, nor did we take extravagant trips. We grew up content with trips to the beach on a Sunday, summers at Boy Scout camp or the Berkeley city camp, and the occasional trip to visit relatives in Seattle, Louisiana, and Texas.

But family isn't the only place where children learn what is and isn't important. The world which we inhabited, much like the world today, screamed of the importance of money. The winners of television game shows were invariably given big piles of money as their prize, and everyone fantasized over having such an opportunity. We knew the names of the richest people in the country, even the world.

What was important about them was their wealth, not what they had done or hadn't done to acquire it. To become a millionaire was to have snatched the brass ring of life, there being nothing higher to aspire to. That was the air we breathed, and we knew no different. Oh, to be a millionaire!

In spite of all that, I chose one of the lowest paying professions there is, that of a professional soldier. For the princely salary of $450.00 per month, I was given responsibility for the lives of forty soldiers in combat. Having said that, neither I nor my wife and children ever wanted for anything. We could afford to eat well, had a decent house to live in, and could afford the occasional vacation trip. And yet, the desirability of becoming a millionaire was still there in the back of my mind.

My big break came in 1981. I was a newly minted major attending the Armed Forces Staff College in Norfolk, Virginia. The class of several hundred officers, drawn from all four of the armed services and our allies, was broken down into smaller groups of about twenty, known as Seminars. The faculty of each seminar was comprised of three officers from different services. Our seminar chair was an Army full colonel, assisted by a Navy commander, and an Air Force lieutenant colonel.

One day, several months into the school, Colonel Hunter, our seminar chair, made the following announcement: "How many of you would like to be a millionaire?"

"Hell yeah", I thought as I shot my hand up. Looking around, it was apparent that my classmates had breathed the same air as me growing up because their hands were up as well.

"Very good. I see that there is an interest. I'm wiling to teach a seminar on how to become a millionaire. Since it's not part of the course curriculum, I will have to teach it outside of normal class hours, probably on a Saturday. How many of you would attend such a seminar?"

Again, the same hands shot up.

"OK. I'll organize it and let you know. It will probably be in a couple of weeks, on a Saturday."

Excited at this unexpected opportunity, I went home and announced to my wife, "Guess what? Colonel Hunter is going to teach a seminar on how to make a million bucks."

"That's wonderful. Can the wives attend?"

"I don't know. He didn't say anything about the wives going."

"When's it going to be?"

"In a couple of weeks, I think."

The following week, as promised, Colonel Hunter announced that he would be teaching his seminar a week hence. I began musing about what I would do with a million dollars. I loved my profession and wouldn't leave it, but being a millionaire in the bargain, how much better could it get than that?

On the prescribed day, we assembled for the seminar, eager to learn the secrets that would make us all independently wealthy.

"Is everyone ready to learn how to become a millionaire?"

"Yes, sir," we echoed in unison.

"OK. Let's get started."

With that, Colonel Hunter shared with us his recipe for personal enrichment.

Every evening after work, he would go home and peruse the stock section of the newspaper, buying and selling stocks. Trading in stocks is a form of gambling and, in order to do that profitably, one has to study the stock market and become knowledgeable about market trends and the performance of individual stocks.

"Hmmm" I thought, "I've never been too interested in the stock market. I rarely, if ever, pay any attention to the performance of the stock I inherited when my mother died. Let's see what's next."

Every weekend, Colonel Hunter would drive around the city and surrounding countryside looking for houses that were neglected, as evidenced by peeling paint, un-mown grass, etc. When he spied such a dwelling, he would knock on the door and ask the owner if they wanted to sell their house. He admitted that this often resulted in rejection but, as he pointed out, if you ask enough people, you eventually achieve success.

Having acquired a house, he would make cosmetic repairs and flip the house at a profit. With profits from flipping houses, one could then purchase an apartment building and start reaping rental income.

"Hmmm," I thought, "when I was with Army recruiting, the thing I hated most was cold calling on the telephone. This involves cold calling at people's homes. I don't think I would be comfortable doing that."

And so it went. It became apparent that Colonel Hunter devoted every moment of his personal time to making money in ways that I would find very distasteful. He had a wife and children, and I found myself wondering when he spent time with them. It appeared that he loved money more than anything else.

Colonel Hunter had done me a huge favor. At the end of his seminar, I was able to say with absolute clarity, "I guess I don't want to be a millionaire."

The notion that being a millionaire was something to be desired above all else was forever put to rest. I also learned that spending time with family and having time to recreate has a value beyond price. I learned the lesson that my parents had taught me by the way they lived their lives, but which was drowned out by the clang-ing, neon messages of society about wealth.

Sufficiency is enough.

Oh, and one other thing, Colonel Hunter had a heart attack. I wasn't surprised when I heard.

26. Awakening

The 1980s were a time of great spiritual growth for me. It was then that I read the Bible from cover to cover for the first time and began to think more deeply about my spiritual life and what God might be asking of me, rather than what I might be asking of God.

In the early to mid 80's, I was stationed at Fort Ord, California with the 7th Infantry Division. Because the terrain at Fort Ord offered limited opportunities for maneuver, most of our training was done at Fort Hunter-Liggett, some 75 miles south of Fort Ord.

Fort Hunter-Liggett is a little-known Army installation which lies 23 miles southwest of the Salinas Valley town of King City. Once the ranch of William Randolph Hearst, it has a tiny garrison but boasts a vast maneuver area perfectly suited to the needs of a light infantry unit such as we were.

Its western boundary is the crest of the Santa Lucia Mountains from which one can look down on the Pacific Ocean. To the north is the rugged Ventana Wilderness. Moving inland from the coastal mountains are rolling hills where magpies sail through groves of coast live oak and broad valleys where cattle and elk roam. Two rivers course through Fort Hunter Liggett, the Nacimiento and the San Antonio.

Closer to the garrison are Mission San Antonio de Padua, the third oldest of the California missions, and William Randolph Hearst's ranch house, The Hacienda, which was designed by the famous architect, Julia Morgan. Standing on the rise where the Hacienda is located and looking over the valley toward the mission, one realizes that the vista is little changed from 1771 when the mission was established.

Prior to the coming of the Spanish, the area was inhabited by the Salinan Indians, many of which still live in the area. There is evidence of their earlier occupation in the form of bedrock mortars, their kiva, or sweat lodge, on the mission grounds and, most strikingly, a painted cave. Their Mount Horeb, the Salinan's holy mountain, is the prominent backdrop to the mission.

Because of its distance from Fort Ord, we would normally deploy to Fort Hunter Liggett for two to three weeks at a time. I grew to love these trips, vowing at one stage that someday I would return as a visitor, a vow I have fulfilled many times over since my return to California in 1995.

On unit deployments, I was kept quite busy performing my duties as battalion Executive Officer or, later, brigade Operations Officer, which limited my ability to savor and enjoy the natural beauty of the place. On other occasions, however, I would go there on reconnaissance with just my driver and perhaps one other officer. It was on these trips I was able to take in the natural splendor of this unique place.

This central coast region teems with wildlife, which is seldom seen when hundreds of soldiers are on hand. But when there were only three of us quietly moving around, nature was less bashful about showing itself.

A mother javelina with her brood in tow, traipsing through the tall grass, their long, hairless tails standing erect and waving back and forth like radio antennae.

A rabbit breaking cover and tearing across my path with a bob-cat in hot pursuit.

A falcon swooping down and grabbing a snake in its talons and then soaring away into the bright blue sky.

The ever-present coyotes trotting across Stony Valley in search of a meal or howling as a pack at night, sounding eerily like an electronic device gone mad.

Cowboys on horseback riding through the brush rounding up cattle.

These are among the timeless scenes that I encountered on my various solo expeditions to Fort Hunter-Liggett. It was like being in a Walt Disney nature film.

Then came the time that my paradise seemed to be in mortal danger. There was to be a major maneuver at Fort Hunter Liggett involving the entire 7th Infantry Division, armored and mechanized forces from the California National Guard, motorized units from the 9th Infantry Division at Fort Lewis, Washington, and even units from the British Territorial Army.

As light infantry, we left a very small footprint on the landscape. We moved by foot and took pains to leave no trace of our having been there. But tanks, armored personnel carriers, and thousands of soldiers?

While on a reconnaissance for our role in this giant maneuver, I stood on a piece of high ground looking down at the San Antonio River valley, I could picture the churned up earth, the smashed trees, the utter destruction that such an exercise could visit on the land. Something awoke inside of me, maybe like a mother's instinct to protect her child. Inside me, something screamed, "No, No, you can't come here, I don't want you to come here."

I was learning. I learned that nature is resilient and can recover from cuts and bruises. More importantly, I learned we are meant to be stewards of the land and to do otherwise is sinful.

27. Two Nations Separated by A Common Language

It is said that the UK and the U.S. are "Two nations separated by a common language."

From 1986—1988, I was assigned on exchange to the British Army where I served as a staff officer at Headquarters United Kingdom Land Forces (UKLF). My rank at the time was Major, and my assistant was Captain Bill Edwards, Parachute Regiment. I was the only American in a rather large headquarters and was, therefore, something of a novelty, particularly my use of colloquialisms.

Wednesday afternoons at UKLF were designated as 'Sports Afternoon' when those so inclined could indulge in sport. I participated in intramural sports and also played for the ULKF rugby side, but, on the afternoon in question, I happened to be in the office. Bill Edwards was to be transferred back to his regiment soon and he was devoting his sporting afternoons to doing forced marches wearing his rucksack.

The telephone rang. "Headquarters, UKLF, Major Sherwood here."

"May I speak with Captain Edwards?"

"He's away right now. Today is sports afternoon and he is out humping."

Once again, the polite silence at the other end of the line alerted me to the potential for misunderstanding but, once again, I held back. Why shatter the delicious image that must have been in the mind of the caller? And why deny myself a good chuckle?

(In the U.S. Army, the activity of marching or moving cross-country with full field equipment is commonly referred to as humping. In the UK and the U.S. civilian world, the term has an entirely different meaning. I have raised some civilian eyebrows when I have spoken of "Humping my ruck").

28. Belize

Early in the morning of New Year's Day, 1986, I stepped off a plane at Heathrow Airport in London to begin a 2 ½ year assignment as an exchange officer with the British Army. There to meet me was my sponsor and soon to be friend, Major Jon Easton, Parachute Regiment. After retrieving my baggage and loading it into a staff car, we headed out to my assignment with Headquarters, United Kingdom Land Forces (UKLF) in Wilton, some 75 miles southwest of London in the county of Wiltshire.

Having flown all night, I dozed off, only to be awakened by Jon who announced to me, "There's Stonehenge."

"Oh, yeah."

Sure enough, there it was, right off the highway. Having flown all night, I was too tired to be appreciative of antiquities. My initial impression was that it wasn't as big or extensive as I had imagined. Being only a twenty-minute drive from where I was to live, there would be ample time to explore it in detail later with my family.

My assignment was as Staff Officer, Grade 2, Overseas Training Exercises, G-3 Training (SO2, OTX, G3-TRG). My responsibility was to mount the out-of-country training exercises for UK-based units, excepting Germany. The British Army conducted several exercises a year, almost exclusively to places that once had been part of the empire. This included battalion-size exercises to

Canada, Kenya, and the U.S., company-size deployments to Australia, Belize, Cyprus, Falkland Islands, Hong Kong and others.

I was expected to travel to all these places and become familiar with them so that I could speak intelligently about them when I briefed deploying units. For someone who craved adventure and travel to new places, this was indeed a dream come true.

One such adventure was to Belize. Belize, formerly British Honduras, had a full-time British garrison, with its headquarters near Belize City, but with outposts on the contested border with Guatemala. In addition to its full-time garrison, Belize also hosted several company-sized deployments from the UK, for which I was responsible.

In October 1987, I travelled to Belize along with Major Martin King, whose company of the Devon and Dorset Regiment was scheduled to train there in the not-too-distant future. Having left RAF Brize-Norton in the UK at a chilly 0530 hours, we arrived in the sweltering heat of Belize at 1300 local time, quite knackered from the trip.

We were greeted by the Chief of Staff, Major Ciaran Snagge, an Australian serving with the Royal Green Jacket Regiment. He, along with his wife, Jenny, would be hosting Martin and me during our visit.

After two days of meetings and briefings with all the key British staff personnel, it was time for Martin and me to get out and see the different training areas. The British Army always included what they termed Adventure Training in all their big exercises, so our first destination was the adventure training site on Caye St. George, an island eight miles out in the Caribbean.

We arrived at the docks in the morning to the sight of a WWII era LCVP, also known as a Higgins Boat, which we presumed was going to be our transport out to the caye. Accompanying the boat was an older looking NCO with a parrot on his shoulder who we assumed was the coxswain. The only problem was, he was staggering drunk.

"That guy is drunk off his ass."

"He looks like a fucking pirate."

"I think he's been out here in the tropics a little too long."

"And he's going to be our driver? I don't know about that."

As it turned out, this was not our transport. We had missed the morning run out to the caye. After a couple of false starts, the adventure training staff sent an orange, motorized inflatable boat, known as the "Rubber Dubby" to pick us up.

Out on the caye, located on a barrier reef, the small training staff introduced us to the adventure training opportunities that would be made available to the soldiers. This consisted primarily of canoeing and snorkeling. At their invitation, we partook of the wares, so to speak, spending a few hours paddling around in canoes and snorkeling on the coral reef.

"Have you ever snorkeled before?"

"Yeah, but never on a reef. There's a whole other world down there. You can see the reef from above, but once you get under water, it just comes to life. I felt like Jacques Cousteau."

The facilities on the caye were quite primitive. The toilet consisted of a short pier over the water about three feet wide and ten feet long, with a small outhouse at the end. Inside the outhouse there was a toilet seat, with nothing below but open air. Everything dropped right down into the water.

"Hey, Martin, you know what they say, you are what you eat."

"Hmmm, I may have to pass on the fish."

The next day, several of us took the "Rubber Dubby" out to Caye Caulker. Caye Caulker is a small island, five miles long by less than a mile wide, approximately 20 miles from Belize City.

Coming ashore after tying up at the small dock in the lagoon, we immediately came to Martinez' Tavern, situated on the edge of what euphemistically would be called the main road around the island.

There were no vehicles on the island and the "road" was nothing more than a sandy walking path. After indicating that we wanted to have lunch, Martinez produced a table and some chairs, which he put right in the middle of the "road". Martinez' specialty, perhaps his only offering, was lobster and chips, washed down with his homemade rum punch. While we sat there enjoying a delicious meal and drinking many bottles of rum punch from reused brown beer bottles, I noticed a middle-aged white man puttering around. He was shabbily dressed in khaki shorts, a light cotton shirt open in the front, a dirty khaki ballcap and bare feet. He didn't look like a tourist.

"What's with the geezer? He doesn't look like a tourist."

"Oh, there's plenty like him around. Mostly Americans. They can live here for next to nothing. The climate is right. There's plenty of food—fruit from the trees and fish from the sea."

"When I was a kid back in the '50's, dropping out and beach combing was quite a thing. As a kid it just meant beach comber pants and a calypso shirt, but I guess it was a real deal. Now I know where a lot of them ended up."

"Well, I suppose there's worse ways of making a living."

"Too right, I think I could get used to it."

That evening, the Snagges hosted a barbecue at their house for Martin and me, as well as several of the permanent staff. Ciaran and his wife Jenny were very sociable and hospitable, making Martin's and my stay most enjoyable.

The next day, Martin and I were taken by helicopter up to Battle Group North, which included a flyover of Holdfast Camp. On the way back we checked out the Guacamayo Bridge over the Macal River. The bridge was nothing more than a concrete viaduct connecting one of the major roads intersecting Belize, which was nothing more than a dirt track.

From there the helicopter took us to Xunantunich, the Mayan ruins on Belize's western border with Guatemala. The site, in pristine condition, was quite isolated and there was no one else there. Martin and I were able to romp around like unleashed Tom Sawyers, climbing the pyramid, and examining the sporting field where the ancient Mayans played a ball game to the death. Having completed our romp in the ruins, we got back on the helicopter and headed back to the main camp outside Belize City. That evening we had dinner with the Snagges at Chateau Caribbean.

The next day, with Major Snagge accompanying us, we headed south, again by helicopter, to check the outposts on the southern border with Guatemala. Departing at 1100 for Battle Group South, we first stopped at Rideau Camp where we had lunch with the officers of 2nd Battalion, the Parachute Regiment. After lunch it was on to Salamanca Camp. All of this was heavily jungled terrain where the Devon and Dorset's would be training on their upcoming exercise.

About halfway back to base, Major Snagge suddenly motioned for the pilot to land. Startled, I at first thought something was wrong. The aircraft seemed to be operating just fine, but Major Snagge clearly wanted to land in this specific spot, an open area adjacent to the jungle. Once we touched down, he jumped out of the aircraft and started walking purposefully toward the jungle.

What the hell's going on? I thought. *Where's he going?*

Not content to sit and wait, I too, jumped out of the chopper and followed Ciaran as he headed for the jungle. About 150 meters inside the jungle, we came upon a huge shed, underneath which was a great Noah's Ark of a boat in the process of construction. Next to the boathouse was a house, clearly hand-made of local materials. A youngish couple, probably in their 30's greeted us. They appeared to be from the U.S. or Canada.

"Hello. What brings you this way?"

"Well, I happened to be in the neighborhood, so to speak, so I thought to drop by and check on the status of my order."

Order, I thought, *what the hell?*

The man, whose name was Kirby, went back to his house and returned with a wooden plate, which appeared to be made of some exotic, local wood.

"I've got most of the plates done. When I finish them, I'll get on to the bowls and utensils. I should be finished in a couple of weeks or so."

"How's the boat coming?"

"Oh, there's still a lot to be done, but we're getting there, as you can see."

"Yes, I do see. It's looking very smart."

After a little more small talk, Ciaran bid them goodbye with a promise to check back in a few more weeks.

Walking back to the chopper, I said, "Who are they and what was that all about?"

"They're what I suppose you would call dropouts. They've been living in that house that they built for a while now, but they decided they would rather go to sea and live on a boat. Hence, the boat they are building. They're harmless. We refer to them as the Eccentric Fringe."

"I've seen a fair amount of that since I've been here."

"Oh, you haven't seen the half of it. This country is full of them. There's a chap who lives in the jungle all by himself. He's friendly enough and will treat you to a roasted iguana should you chance to visit him."

"So, what's your connection with them?"

"To make a little money, they turn things on a lathe. They are making me a set of plates, a big salad bowl, and individual salad bowls."

"But there's no electricity."

"Pedal power, mate, pedal power."

"Wow, now I've seen everything."

But I hadn't seen everything, not by a long shot.

That evening back at the Snagge's, I was talking to Ciaran's wife Jenny.

"What kind of crafts is Belize known for? Whenever I go somewhere, I try to bring my kids something from the local culture."

"Well, I don't really know. Nothing immediately comes to mind. Perhaps wood carving."

"That would be good. Where do you think I could find some?"

"Well, there's always the airport, but I've heard that you can get wood carvings at the prison."

"The prison!?"

"Yes. Apparently, the prisoners are allowed to do wood carving and offer their work for sale."

"That sounds fantastic, what a hoot."

"Let me look into it."

I had a personal stereotype that Australians were adventurous. Everything about Jenny Stagge reinforced that stereotype. She seemed game for anything and the more exotic, the better. By mid-morning the following day, she had completed her research into the matter, enthusiastically announcing, "The prison does, indeed, sell wood carving. The carvings are done by the prisoners.

They are allowed to sell their wares in order to make a little money. I think it would be a cracking idea to pay them a visit."

"Wow, that sounds great. I'd love to check that out."

"I want to see it as well. Shall we go down there after lunch?"

"Absolutely. That sounds perfect."

So, after lunch, we drove down to the prison, appropriately located on Gaol Lane. It was right in the center of town, not far from the harbor. From the outside, all that was visible was a high, dun-colored wall which surrounded the entire block, topped with barbed wire. The arched entrance consisted of a pair of huge wooden doors with masonry towers on either side. When opened, a bus could easily have driven through. Embedded within the left door was a smaller door, clearly designed as a personnel entrance. There was no one around and no instructions posted about gaining entry. We began pounding on the door and calling out.

After getting no response, we were about to conclude that our plan was not to be. All of a sudden, a face appeared in a peep hatch in the door.

"We want to buy wood carvings."

"Carvings?"

"Yes, wood carvings."

"Wait."

He closed the peep hatch and disappeared from view. A few moments later, the personnel door opened and a man in a sweat-stained tan uniform beckoned us to enter. Stepping through the door, our host directed us into the guard's office immediately to the left of the entryway. The office was large, with a few desks around the perimeter. One wall was the outer prison wall, the other three walls had windows with a view of the prison yard, where the prisoners were milling around.

We again explained that we were interested in buying prisoner's wood carvings. Told to wait, there was a quick conversation among the guards, and then one of them exited onto the yard.

Pretty soon, prisoners holding paper bags started congregating around the office windows and peering in. One, in particular, was pressing his face against the window and mouthing something. Then, the door was opened, and the prisoners filed into the office, sat on the floor, and began to unpack their bags and display their wares.

It was amazing. Like an open-air market, each vendor was doing everything possible to gain the attention of Jenny and me. There were bulls, parrots, and other tropical birds and…spiders. The prisoner who had his face pasted against the window revealed what he had been mouthing when I first saw him.

"Spiders! Spiders!"

Against a spider web, he had carved a large, tarantula-like spider perched on the web, each delicate leg intricately carved.

How to choose? The experience was so unique and each craftsman so deserving, I would have liked to buy it all. But choices had to be made. I finally settled on the spider. It was unlike anything I had ever seen, and I thought it would be the most appealing to my kids. Also, the sales technique of the craftsman was not to be discounted.

Business concluded, Jenny and I thanked all. The prisoners packed up their wares and returned to the yard, and the guards escorted us to the door, biding us farewell.

As we drove back to the Snagge's, Jenny commented, "That was an interesting experience."

"You can say that again! That was worth the price of admission. Think of it. In addition to local crafts, what a story to go along with it!"

Thus ended my reconnaissance to Belize. In the eyes of many

Americans, the trip would have been viewed as a complete boon-doggle. But the fact was that I accomplished everything my superiors wanted me to. I met all the key players on site. I visited and became familiar with the base facility and all training areas, and I was prepared to brief deploying units based on first-hand knowledge. The fact that I had an interesting and fun time was beside the point.

Major Snagge heading into the jungle

Major Martin King on pyramid at Xunantunich, Belize

L-R: Major Ciaran Snagge, Royal Green Jacket Regiment and
Major Martin King, Devon and Dorset Regiment, Belize

Belize Prison

29. The Command Sergeant Major Meets His Match

One of the advantages of an all-volunteer military is that it is able to release back to civilian life those volunteers for whom military service is not the right fit. After all, what professional team would want to hang on to a player who didn't want to be there, and wasn't willing to play to the best of their ability?

Smitty was just such a soldier. He undoubtedly enlisted with the best of intentions but found himself unable to adapt to army life. He wasn't a bad sort; he just wasn't cut out to be a soldier. While awaiting his discharge, Smitty was working in the S-1 section (personnel administration) at the battalion headquarters, answering the phone and performing errands.

Southern "Buddy" Hewitt was cut out to be a soldier. Enlisting in 1955 at the age of seventeen, he was as hard-boiled and tough as any soldier that ever wore combat boots. He had been wounded three times and had participated in some of the most brutal battles of the Vietnam War. Now, in the twilight of his career, he was the Command Sergeant Major of the 10th Mountain Division, perhaps the best and toughest division in the Army at that time. He was known, feared, and respected by every soldier in the division.

One day, everyone but Smitty was away from the headquarters at lunch, he being left to answer the telephone and greet any visitors. CSM Hewitt picked that moment to make an unannounced visit to the battalion.

His sinewy, compact body was encased in starched camouflage and gleaming boots. His somewhat menacing face, creased and brown from much time spent outdoors, stared intently at Smitty as he barked, "WHO'S THE NCOIC[7] HERE!?"

Unruffled and without even bothering to stand up, Smitty casually replied, "I don't know what NCO you see, Sergeant Major, but you're the only NCO I see."

Knowing that voicing the rage that he felt would only seem petty and knowing that he had been bested, CSM Hewitt did the only thing he could do. He retreated, perhaps for the first time in his life. Turning on his heel, he stormed out of the headquarters.

(This story would have been lost to posterity but for the fact that there was a witness. Lieutenant Mark "Duke" Ellington had come to the headquarters to make some copies and was able to witness the whole affair from his concealed position in the copy room. Lieutenant Ellington was one of those who appreciated the humor of life and was barely able to keep from laughing out loud as this drama unfolded).

---

7. NCOIC—Non-commissioned Officer In Charge

30. Montaneros Bravo

"How would you like to take your Battalion to Honduras?"

"Hell, yes, sir. When?"

"We're looking at a deployment date in early November."

This conversation took place between my brigade commander, Colonel Wolf Kutter, and me in September 1988 as we walked between his headquarters and mine. I had just taken command of the 2nd Battalion, 22nd Infantry Regiment, the Triple Deuce, of the 10th Mountain Division at Fort Drum, New York in August, and here I was being given the opportunity to deploy out of country with my battalion. Any deployment was welcome, but to go to an exotic, semi-dangerous location...well, it just didn't get much better.

Planning for this deployment would take place simultaneous with our normal, vigorous training schedule, so we were quite busy. We would be going as a battalion combat team, accompanied by an artillery battery, a platoon of combat engineers, as well as various service support personnel, which meant that the leaders of these various components would have to be a part of the planning process.

There was much coordination to be done with higher echelons of the Army, as well as the Air Force.

My two years as an exchange officer in the British Army had

taught me a lot about overseas deployments, one of which was the importance of incorporating humanitarian aid projects into the overall scheme. This was the decent thing to do, plus it had the practical effect of presenting the United States as a friend rather than an intruder. Accordingly, I set my chaplain, Captain Dave Scheider, to scouring Upstate New York for donations of clothing and books written in Spanish. Building projects and medical assistance would have to be determined on the ground after consultation with the local population.

We were to have a company from the Honduran 10th Mountain Battalion attached to us. This was an exciting prospect, but also presented the challenges of liaison with a foreign nation, language differences, and the need to identify Spanish speakers within the battalion who could act as interpreters.

In addition to the logistics of getting the battalion combat team from Fort Drum to Honduras, it was necessary to plan the operation on the ground. This required a reconnaissance.

In October, I, along with my Operations Officer (S-3), Major Wade Roberts, and my Supply Officer (S-4) Captain Dale Goble, flew to Honduras on a commercial airliner, landing at Tegucigalpa, the capitol city. Our base of operations would be Soto Cano Airbase, located on the western outskirts of Comayagua.

My initial intent for the training was to focus on live-fire exercises. Realistic live fire training was heavily emphasized in the 10th Mountain Division, and rightly so. Half of our training, both day and night, was to be done using live ammunition. A battalion that had deployed to Honduras before us had focused on live firing and I was pretty much of the same mindset. I also wanted to operate as close to the Nicaraguan border as possible. President Reagan had identified the Sandinista government as a threat to the United States, and I wanted to get as close to the action as possible.

After meetings with Honduran officials and the U.S. Army colonel in charge of operations in Honduras, it became clear that I had great flexibility in choosing where I was going to train and what type of training I was going to do—within certain parameters. There was no way we were going to be allowed close enough to the Nicaraguan border to mix it up with the Sandinistas.

As far as live fires went, there were places we could do live fires, but there were concerns about not tearing up the habitat. I quickly came to some conclusions and made some decisions.

"Wade, we can do better, more realistic live fires back at Drum. There's no point in coming all the way down here to do something we can do better at home. Let's see what else they have to offer."

After inquiring where else we might train, it was suggested that we look at a mountainous area to the southwest, between Comayagua and San Pedro de Tutule. The border with El Salvador, where a civil war was raging, was about thirty miles further on.

Heading south from Soto Cano, we left the paved road and connected with a small, unpaved track that carried us up into a mountainous area, forested at the upper elevations with pine trees but jungle-like where streams cut through the valley below. There were some small villages in the valley and some individual habitations scattered around in the mountains.

"Wade, this is perfect. This is exactly the kind of terrain for light infantry to operate in. I want to run a low-intensity scenario against an organized guerilla force like the Sandinistas. We'll do a force on force, with one company as the "G's" and the other two companies as U.S. maneuver elements. I won't have a reserve, but this will ensure that everyone will get into the action.

"Dale, I want you to see about hiring mules. There're no roads here so re-supplying by vehicle is out of the question. We'll have helicopter support, but I want to experiment with alternate means of transport and re-supply. In addition to using mules for re-supply, I want to use mules to transport the 81-mortar platoon. Humping those bad boys in this terrain would be a killer."

"Yes, sir, I'll see what I can do."

Dale was able to hire enough mules to handle the mortar platoon as well as resupply of rations and water.

Warfare is called both an art and a science. The scientific part involves understanding how to move and supply the troops as well as understanding weapons systems, their capabilities and employment. The artistic part involves leadership and concept of the operation.

There was much more coordination to do with the Honduran Army and our logistical base at Soto Cano, but the most important thing, the Concept of the Operation had been decided on. Going forward, all coordination would be done in furtherance of that concept.

Back at Fort Drum following the reconnaissance, the final planning and preparation proceeded at a furious pace. Not least among the planning was the risk assessment. Every possible risk to the safety of the soldiers had to be considered and addressed — weather, terrain, dangerous wildlife, etc. Good training needs to be realistic and challenging, necessitating to the degree possible the mitigation of risk.

For example, it was determined by someone much higher than me that this mission involved the possibility of an encounter with a hostile force and that we should carry live ammunition to counter such a threat. Mixing live ammunition with blanks was fraught with risk. To mitigate this risk, it was decided that soldiers would carry their live ammo in their left breast pocket and that blank ammo would be carried in the normal ammo pouches in the combat vest.

On 6 November 1988, a cold, drizzly day in the North Country of New York, Task Force Triple Deuce was transported from Fort Drum to Griffiss Air Force Base, where we boarded planes for the long flight to Honduras. Having left Fort Drum, where the daily temperatures were typically below freezing, we arrived in the tropical heat of Honduras. The troops were extremely fit and well trained in the need to hydrate, but it was still necessary to spend a couple of days at Soto Cano getting acclimated and preparing for the upcoming operation.

One platoon, 1st Platoon, B Company, under the leadership of First Lieutenant Allen Zick and Sergeant First Class Hutton was immediately dispatched to the Honduran 10th Mountain Battalion to train with them and to teach them how to use night vision devices and other U.S. military technology which was new to them. Lieutenant Zick and his platoon would remain with the Hondurans until the beginning of the exercises when he and the Honduran soldiers would join the battalion.

Soon after our arrival, I started hearing a lot about Bimbo Bingo. "What the hell is Bimbo Bingo?"

"Well, sir, once a week the local women are allowed on base to play bingo. But before they play bingo, there is a dance. You should check it out."

"Damn right! I've gotta see this."

Bimbo Bingo being the following evening, I went up there with my interpreter, a soldier in the battalion who happened to be Honduran. Sure enough, at the prescribed time, the gates were opened to a flood of local damsels, all decked out in their best party clothes, who crowded into the enclosure where the dance and subsequent bingo were to take place.

Not only was my interpreter Honduran, he also seemed to have family connections in the local area. "Sir, my cousin wants to dance with you."

"Your cousin!?"

"Yes, sir. That's her there," he said, indicating a short, chubby young woman standing not six feet away.

Never much inclined to dancing, I would normally have declined, but it seemed as though it would have been very rude to do so, so off I went with the cousin, stepping it out to a salsa number. Mercifully for me, the dance portion of the festivities didn't last very long, abruptly stopping at a prescribed time unknown to me.

It turned out, to my surprise, that playing bingo was far more exciting to the ladies than dancing. I suppose that was because they could listen to music and dance any time they wanted, but organized bingo was something else. I thanked my dance partner and made my exit, much amused.

While some of us were being entertained by Bimbo Bingo, the first phase of the operation was getting underway with the clandestine infiltration of the Scout Platoon into our Area of Operations (AO) on foot. Not long into their move, the platoon leader radioed in with a report that the heat and dense vegetation was making it difficult to make any progress. My instinct was that with additional experience in this part of the world, this seemingly impossible mission would be achievable.

However, the objective here was to insert the platoon into the AO, not grind them down by teaching them a hard lesson. I told them to return to base while I came up with an alternate plan. To my XO and S-3, Majors Jeff Spara and Wade Roberts, I said, "Line up a school bus. After it gets dark, we'll put the Scouts on it and have them get down on the floor where they can't be seen from the outside. Pick a remote spot on the edge of the AO where they can dismount and infiltrate. We may even gain some time this way and they will be fresh when they get there."

After the Scouts were established, the remainder of the battalion task force was inserted. A Company, under the command of Captain Steve Fields, was on the western edge of the AO. Their mission was to play the role of the enemy guerilla force. They were to vigorously resist the advance of the other two rifle companies and enlist the help of local villagers in their efforts.

B Company, under Captain Chip Quade, set up in the northeast part of the AO, and C Company, commander by Captain Ron Corkran, in the southeast. B and C companies were to sweep west along parallel mountainous ridgelines, separated by a river that ran through a deep, jungle ravine.

The battalion command post (CP) was set up atop a ridge at the southeast corner of the AO. Nearby, the battalion aid station was set up under the supervision of Warrant Officer (WO) Dougan, the battalion's Physician Assistant (PA). In essence, he was the Battalion Surgeon, capable of performing most of the medical treatment that would be required in an infantry battalion.

"Chief" Dougan was a character among characters. He had been a Special Forces medic during the war in Vietnam, later qualifying as a PA. He had a quirky personality and sense of humor, and was much liked and respected throughout the Triple Deuce.

Word of our presence spread very quickly, and it wasn't long before we had a steady stream of customers at our aid station — almost exclusively mothers and their infants.

"News seems to travel fast around here."

"Roger that. Once word gets around that the Americans are here, they know we'll have medics with us."

"They probably don't have access to any kind of health care."

"This is amazing. We bitch if we have to park on the far side of the parking lot and here these people walk who knows how many miles through the mountains, carrying their babies, in the hopes that we can help them. Amazing!"

"Mister Dougan! Whatcha got goin' on over there?"

WO Dougan was holding a crying infant in his arms and sporting a big grin. In the normally hard-edged world of the infantry, it was both incongruous and pleasing to see this hard-boiled career soldier cradling an infant and witnessing the complete trust in the mother's face that we were there to help.

"We have the best free clinic in town. You know we aim to please. This is my star patient."

This was a humanitarian aid project we hadn't even anticipated. We had planned to have the medics visit the tiny villages in the area, but we hadn't expected them to come to us.

On the ridge about forty meters beyond the aid station, there was a small, cleared area in which was a small, one-room mud hut with a thatched roof. Next to it were a few rows of scraggly corn and some scrawny looking chickens.

The day after we arrived, a little boy of about five years came from there and presented us with a stack of fresh, still warm tortillas, a gift from his mother. To refuse a gift would be an insult, so the gift was graciously received, and eaten with relish as a welcome change from combat rations.

"Bloody hell! She doesn't have a pot to piss in and she's sending us a gift of food? I can't believe the generosity of these people. Let's make sure and load her up with MRE's." (Meals Ready to Eat—our combat rations).

"Roger that, sir. We have plenty."

Later, I wanted to convey my appreciation of the Honduran people to the Honduran captain attached to us. He was a Comandante X type, with tailored uniform, an arrogant face, and always sporting mirror-lens aviator sunglasses.

"Your people are really beautiful," I said.

I couldn't see his eyes, but I could imagine the disdainful look in them. After a pause, he replied, "Some of them are."

I immediately realized that, in this country, the army wasn't the protector of the people, the people were their enemy.

While the rifle companies were settling in and doing their planning, there was a lot of activities in other quarters. Helicopters were delivering the 105 mm howitzers to the battery firing position, situated where they could provide fire support to the whole battalion. The sapper platoon (combat engineers) was beginning their project of building a footbridge over a stream so that, during the wet season, the inhabitants of the mountainous area could make their way down and back without having to wade across the stream.

Captain Dale Goble and Lieutenant "Pee Wee" Pieragostini had each taken over a string of mules, with accompanying Honduran muleskinners. Captain Goble would be overseeing resupply by mule and Lieutenant Pieragostini, the 81mm Mortar Platoon leader, would be moving his mortars through the mountains by four-legged transport. Back at Soto Cano, the XO, Major Jeff Spara, was making sure that we received continuous logistical support.

Before the operation could kick off, we had to receive the company of Honduran soldiers and integrate them into our maneuver elements. Along with them, we had to retrieve Lieutenant Zick's platoon and return them to B Company.

Major Roberts organized a small patrol to accompany us and provide security as we moved to the landing zone (LZ), a few kilometers away, where the Hondurans would be inserted by Blackhawk helicopter. Moving downhill, it didn't take long for us to arrive at the LZ, a large, open, relatively flat area amid the forested slopes of the mountain.

While we waited in the woods at the edge of the LZ, the Blackhawks started arriving. As each bird touched down, the soldiers jumped off, weapon in one hand, rucksack in the other, ran about fifteen meters, then flopped down in the prone position behind their pack, weapon at the ready. Once the birds lifted off, they jumped up and hustled off the LZ to their assembly area, making room for the next flight to land. Almost all were short, muscular young men who moved with the ease and certainty of trained soldiers.

Soon, Lieutenant Zick's platoon started arriving. Sergeant First Class (SFC) Hutton, the platoon sergeant, was on the first bird. As the soldiers moved to their assembly area, SFC Hutton joined Major Roberts and me in the wood line.

"How was the experience with the Hondurans?"

"It was something else, sir. The first day we were here they killed a boa, a big one. I wouldn't go near the water after that. They like having the boas around, though, because they kill off the poisonous snakes and rats."

"What are the troops like?"

"Sir, if they come on our people, I feel sorry for our people."

"Really? How so?"

"Sir, these people can run through these mountains. We were at high altitude, and they just run up them like it's nothing. I thought we were in really good shape, but we were crapped out after a click."

We talked on for a while until LT Zick arrived on the last bird and joined us.

"What did you think of their soldiers?"

"Well, sir, their officers and NCO's know what they are doing, but the soldiers are new and are learning. They are all from around here. They know the terrain very well and can hump their asses off."

"What else?"

"They love the night vision goggles. When we were training with them at night, I could hear some of them saying, 'milagro, milagro.'"

We talked on for a while about the Honduran tactics and methods of communication, and then LT Zick and SFC Hutton returned to their platoon. At the same time, the Honduran soldiers were being integrated into the task force.

Once the Hondurans were integrated into B and C Co., the operation commenced with each company advancing within their assigned sector. A Company, under the command of Captain Steve Fields, was the "enemy" guerilla force opposing B and C Companies. I had instructed him to make sure that his soldiers engaged the B and C Company soldiers as much as possible. He did his utmost to accomplish that, thus giving all soldiers the opportunity to practice their small unit tactics and individual combat skills.

The main CP remained where we had originally set up, under the control of Captain Scott McBride and Master Sergeant Dolan. The tactical CP, consisting of myself, Major Roberts the artillery liaison officer, and various radio operators, followed along behind C Company Command. Control of dispersed units in rugged and mountainous terrain was a challenge, but in a good way. Only through challenge does a unit and all its members grow and develop.

I continued to be struck by the primitiveness of our surroundings. One day, while halted in a concealed position on a mountainside, we were treated to yet another reminder that we weren't back at Fort Drum.

"Wow, look at that, will ya?"

"Damn! Pancho Villa will be along any minute."

Riding along a narrow mountain path on the ridge opposite us was a man on horseback wearing a cowboy-type hat, nondescript clothes, and sandals. A rifle was strapped across his back and a large machete hung at his waist. He didn't look like someone to mess with.

"Do you think he sees us?"

"I don't know, but if he does, it doesn't seem to concern him."

Elsewhere, the Catholic chaplain who accompanied us was saying mass in one of the small villages at the lower elevations. The area was so remote that there was no regular priest or even a church building. A lay woman was the local spiritual leader, and it was she who organized the locals for the outdoor mass. The chaplain said the mass in Spanish, but there were also several English-speaking soldiers in attendance.

Since the Catholic mass follows the same format no matter where one is in the world, everyone was saying the prayers in their own language, creating a cacophony of sound. Nevertheless, it was a beautiful opportunity for the soldiers and locals to join together as equals.

At the field trains, the Honduran muleskinners were loading the mules with supplies, while simultaneously teaching our soldiers how to do it. Their equipment was very primitive: saddles made of wood cross-pieces, cinches and lashings of rough hemp rope rather than leather straps; yet they knew how to make it work. Heavy boxes of 81 mm mortar ammunition, 5-gallon water cans, cases of MRE's, the mortar tubes, base plates and bi-pods were all loaded.

Once the loading was complete, the caravan moved out along narrow mountain paths. Captain Goble, the Supply Officer, and the Mortar Platoon Leader, Lieutenant Pieragostini, or Pee Wee as he was known, did a fantastic job of learning on the job everything there was to know about transport by mule.

While the tactical operation was ongoing, the aid station went mobile, making house calls in the various tiny villages and administering first aid and other medical treatment as they were able. In one village, Staff Sergeant (SSG) Toste and another medic treated a woman with a huge open wound on her leg. As big as a fist, the flesh was eaten away right down to the bone. "Ask her how long she has had this."

"Senora, cuanto tiempo llevas con esta herida?"

"Diesciseis anos."

The woman further explained to SSG Toste that there were no doctors available and no explanation for why this wound wouldn't heal.

While one medic carefully cleansed the wound with hydrogen peroxide and then dressed it, SSG Toste explained to her how to change the dressing, leaving her with a generous supply of gauze bandages and wrapping tape.

"Fuck me! It's hard to imagine having something like that and just having to live with it, with no hope of ever getting healed."

"Welcome to the rest of the world."

After several days of B and C Companies advancing through their sectors and engaging the "guerrillas", the tactical exercise closed in on one of the larger villages on the eastern edge of our Area of Operations (AO). It was here that the senior commanders in the Honduran Army wanted to deliver the clothing and other goods that Chaplain Dave Scheider had collected. They wanted to make a grand spectacle of it for their own political gain.

I had seen this coming and had directed Dave to clandestinely take half the stuff to the town of La Paz and give it to the parish priest. Such was the cynicism I had developed toward the Honduran power structure, a significant portion of which rested with the Honduran military.

The ceremony was to be presided over by several high-ranking Honduran colonels, who arrived by helicopter, blowing the tin roof off one of the village buildings in the process. After several speeches, to include one by the U.S. Army colonel in charge of U.S. forces at Soto Cano, the goods were delivered to the locals, amid much emotional praising of God for this bounty.

As the exercise ended, the units assembled to await transportation back to Soto Cano. All seemed to have had a memorable experience, but were glad it was over. There was much storytelling, banter and laughter—all signs of a positive experience.

Every soldier who participated in Operation Montaneros Bravo would have their own stories to tell, depending on where they were and what they did. For all, it is safe to say that this experience was one of the highlights of their time as soldiers. It was for me.

THE HONOR OF THE REGIMENT

Back at Soto Cano, it would take some days of cleaning up and getting organized for our re-deployment to Fort Drum. While we had been in the mountains, a battalion of Marines had been conducting an operation on Isle El Tigre, or Tiger Island, located in Fonseca Gulf on the south coast of Honduras. They were at Soto Cano at the same time as Triple Deuce, waiting for their re-deployment to their home station.

On Soto Cano was a large service club where the soldiers and Marines could unwind in the evening. Mixing Marines and Mountaineers was a volatile mixture, especially when alcohol was added. Accordingly, Command Sergeant Major (CSM) Claus Madsen and I decided to be on hand at the club to make sure that there was no unpleasant drama. The Marines obviously had the same idea, the Marine Regimental Commander (a full colonel) and his CSM being there to keep an eye on their flock.

Our collective instincts were correct, as there was a palpable tension between the Marines and Mountaineers of the Triple Deuce.

After introducing ourselves and exchanging some pleasantries, the Marine colonel challenged CSM Madsen and me to a game of pool against him and his CSM. Of course, I accepted the challenge but was simultaneously aware that now the honor of the regiment rested on the pool skills of CSM Madsen and me.

I had no idea of CSM Madsen's skill, but I knew that I hadn't played much pool since my wild days in the 8th and 9th grade when my cronies and I would hang around the pool room at Golden Gate Lanes, shooting pool and smoking cigarettes for hours on end. Also, I had no idea of the skills of our competitors. Were they pool sharks or were they amateurs like us?

"*Oh well*", I thought, "*you pay your nickel and take your chances.*"

"Stripes and solids OK?"

"Yes, sir, sounds good to me."

"Since we challenged, you want to break?"

"Sure thing, sir."

The balls were racked up and I gave it a mighty break, hoping to sink one and get off to a good start. Nothing went down.

The colonel stepped up and made an easy shot, sinking a solid. He missed his next shot, which passed the play to CSM Madsen. He sunk a stripe to put us into play, but missed his next shot.

By this time, the game was attracting the attention of Marines and Mountaineers, which added to the pressure to do well. This was no longer a quiet game between four gentlemen, now it was the Army versus the Marines.

The Marine CSM took his turn, sinking another solid. Then, on his second shot, he sank another.

"Fuck, is he going to run the table?" I wondered.

But he missed his third shot, much to my relief. We were one ball down, but still in the game.

Back and forth it went, shot for shot and it became apparent that we were evenly matched. This was not the first time our opponents had played pool, but they were by no means pros.

We were still one ball behind when the Marines cleared the table of solids, leaving only one stripe and the Eight Ball on the table. To win, they had to call the shot and sink the Eight Ball. It was the Marine CSM's shot.

"Eight Ball, side pocket," he announced, pointing at the pocket with his cue.

He shot and missed.

Now it was my turn to sink the remaining stripe and even the game. The only shot was a bank shot which I didn't think I could make. I walked around the table examining the trajectories, trying to visualize not only the path of the cue ball, but also where it needed to strike our ball.

Not for the first time in my life, "The Force" was with me and I made the shot.

"Holy shit," I thought, "the game's on me."

The cue ball sat approximately where the ball is placed for the break and the eight ball was about six inches in front of the corner pocket, a seemingly easy shot. But I knew from experience that there were hazards here. A hard shot will usually sink the target ball, but often the cue ball follows right behind, resulting in a scratch and loss of the game. Too hard a shot can also cause a table scratch if the ball flies off the table. A soft shot can result in the target ball meandering off course and missing the pocket.

I was experienced enough to know the hazards, but not enough to know which shot was best. I opted for a soft, but firm shot, aiming low on the cue ball to give it some back spin, hopefully drawing it back after the strike.

"Eight Ball, corner pocket," I announced, pointing to the pocket with my cue.

After powdering my hands, chalking my cue, and taking a couple of practice strokes, I made my shot and the Eight Ball rolled smoothly into the pocket, the cue ball resting where the strike had taken place.

A mighty cheer went up from the spectators and there were handshakes with our opponents.

More importantly, this seemed to break the tension that had existed between the Marines and Mountaineers. Good will permeated the room as Marines and Mountaineers took to dancing, solo or with one another, stomping around in their boots like drunken clog dancers.

Most importantly, the honor of the Triple Deuce was upheld.

3d Platoon, A Co., 2nd Bn, 22nd Inf. Mountains in which
the operation took place are in the background.

LT Vinny O'Neil, LT Stu Lyons, CPT Ron Corkran, LT Rodney Pitts, LT Shawn Williams, of C Co., 2d Bn, 22d Inf "Triple Deuce, and LT Trent Andrews Fire Support Officer, C Battery, 1st Bn, 7th Field Artillery; Soto Cano Airbase, Honduras. Mountains can be seen in background.

31. Sic Semper Salaco

In May of 1989, while I was commanding the 2nd Battal- ion, 22nd Infantry Regiment, affectionately known as the Triple Deuce, we were deployed from our home base of Fort Drum, New York to Camp Le Jeune, North Carolina to participate in a major military exercise code named Solid Shield.

Camp Lejeune, a Marine Corps base located on the southern coast of North Carolina, annually played host to this exercise, which engaged all of the military services in mock warfare. The Navy, located off shore, would provide naval gunfire support as well as launch the Marines, who would land on the beach. Paratroopers from Fort Bragg would jump in, and helicopter-borne soldiers from Fort Campbell would fly in. The battle worthiness of these units would be tested, as well as their ability to operate effectively with one another.

The Triple Deuce, along with a couple of Marine battalions, were organized into a provisional brigade under the command of Marine Colonel Kurt Schreiber and given the task of playing the enemy force against this onslaught of U.S. military might. In spite of overwhelming odds, the Triple Deuce gave a good account of itself and earned the praise of everyone up the chain of command, from Colonel Schreiber to Admiral Rickman of the Atlantic Fleet.

Following the exercise, we were billeted on Camp Le Jeune

as we cleaned up ourselves and our equipment in preparation for return to Fort Drum and our parent unit, the 10th Mountain Division. Satisfied that everything was on track, I summoned the battalion adjutant, Lieutenant Mark Pieragostini, "PeeWee, I want you to organize an officer's call for this evening, right after chow. Find out where the nearest O-club is and put out the word to the company commanders."

"Yes, sir, will do."

It turned out that there was an Officer's Club annex within walking distance of our billets. After the evening meal in the mess hall, we assembled and walked to the club.

Entering the dimly lit bar, I noted that the 1960's décor was typical of all military clubs — chrome chairs and stools with Naugahyde upholstery. The tabletops and bar were of Masonite, the bar also sporting a padded Naugahyde bumper. It was tasteless but functional.

I was accompanied by Major Jeff Spara, the Executive Officer, and Major Wade Roberts, the Operations Officer, to whom I said, "Grab us a table over there and I'll get some beers."

We older guys settled down at our table while the younger officers gathered in small clusters closer to the bar, presumably recounting their exploits of the preceding days. The atmosphere was decidedly low key and mellow.

"Wow, everyone's pretty laid back."

"Yeah, I think they're tired."

"I am, too. That was a great exercise, but I am definitely whooped out."

The bubble of serenity was punctured with the arrival of a group of helicopter pilots from the 101st Airborne Division. They were obnoxiously loud, and it wasn't long before their attention was drawn to the young officers of the Triple Deuce standing close by.

"Hey, what do we have here? Where's your 101st patch? I never even heard of your outfit. You're not even aviators."

Sitting on the sidelines watching this go on, I said, "Look at those assholes. I hate fuckin' aviators."

"Yeah, me too. They're right up there with MP's."

My parental instincts coupled with my dislike for aviators made me want to go over and intervene. My wiser instincts told me that my boys could take care of themselves and that it would only shame them if I stepped in.

One aviator in particular seemed to relish the role of bully as he zeroed in on Lieutenant Fred Johnson. Poking Fred in the chest with his index finger, he kept saying, "Where's your aviator wings? Where's your wings?"

The next thing I heard was a loud and resounding "Thwack," followed by loud shrieking.

"He hit me! He hit me!"

The aviator was screaming like a little boy who had just been spanked, which, in a sense, he had.

Now I was on my feet, fearing that it might turn into a brawl, but all of a sudden the bravado had leaked out of the 101st boys. I stepped up to the wounded aviator, who was still shrieking and holding his hand to his face where Fred had slapped, not punched, him.

"Quit acting like a baby. What did you think was going to happen?"

"Call the MP's," he said to one of his chums as he turned away from me. Sure enough, a few minutes later the MP's showed up and took a report on the incident.

"Do you believe that? They called the MP's. What a bunch of pussies!"

The MP's didn't detain anyone, but now that it was reported,

I needed to make sure that Colonel Schreiber heard it from me before getting it from another source. Not wanting to disturb him at night, I decided to call him first thing in the morning.

Colonel Schreiber had been a great commander to serve under and I feared spoiling it all by bringing him this bad news, but bad news doesn't improve with age, so I had to get on with it.

"Good morning, sir. We had a little incident last night at the Officer's Club that you should know about. Some aviators from the 101st were giving my lieutenants a hard time and one of my lieutenants smacked one of them."

With the hard edges of a New York accent, Colonel Schreiber replied, "I applaud you, Colonel Sherwood. I'm sure they got whatever they deserved. Why don't you report to my office at zero nine hundred and we'll take care of this."

Hugely relieved by his reaction, I said, "Yes, sir" and hung up the telephone.

Promptly at nine o'clock I reported to Colonel Schreiber. We sat and chatted for a few minutes, during which he allowed that he had a distaste for aviators and their arrogance. Then he suggested that we take a walk down to meet with the Provost Marshal, who would be the recipient of the MP report and the person who would decide what would be done about it.

The Provost Marshal seemed to be expecting us when we arrived, and it also seemed apparent that he and Colonel Schreiber had known each other for a long time. We had a cup of coffee and made small talk, at the conclusion of which the colonel said to me, "You've been before the Provost Marshal, the matter has been discussed, and it is resolved."

Colonel Schreiber dismissed me while he stayed on to reminisce for a while longer with his old pal.

"Sic semper salaco"—"Thus ever unto bullies."

Byrne N. "Buzz" Sherwood was commissioned as a Second Lieutenant of Infantry in 1970 upon his graduation from LSU. During a twenty-three year career in the U.S. Army, he commanded infantry units at platoon, company and battalion level. Following retirement from the Army, he taught history for ten years at Richmond High School in Richmond, CA. Now fully retired, he divides his time between his local parish church, ministering to incarcerated youth as a volunteer chaplain at the Alameda County Juvenile Detention Center, as a member of the Contra Costa County Search and Rescue Team and as an officer in VFW Post 8063. His hobbies include hiking, writing and music. He and his wife, Diana DeGracia, reside in Lafayette, CA.

www.ingramcontent.com/pod-product-compliance
Lightning Source LLC
Chambersburg PA
CBHW031457120626
46545CB00005B/1641